MEP 804A/B AND 814A/B 15 KW GENERATOR SET MANUAL TM 9-6115-643-10

GENERATOR SET
SKID MOUNTED
TACTICAL QUIET
15 KW, 50/60 HZ, MEP-804A
(NSN: 6115-01-274-7388) (EIC: VG4)
15 KW, 50/60 HZ, MEP-804B
(NSN: 6115-01-530-1458) (EIC: N/A)
AND
GENERATOR SET
SKID MOUNTED
TACTICAL QUIET
15 KW, 400 HZ, MEP-814A
(NSN: 6115-01-274-7393) (EIC: VN4)
15 KW, 400 HZ, MEP-814B
(NSN: 6115-01-529-9494) (EIC: N/A)

edited by
Brian Greul

The MEP series of Military Generators are reknowned for their quiet, durable operation and conservative power ratings. This is the operators manual for the 15KW version of the generator issued under models 804 and 814. The A series has analog controls and the B series has digital controls. Various units are manufactured for the US Government by different contractors with different power plants. This book is a reprint of the operator manual published by the US Army. It is printed as a courtesy to enthusiasts and owners of these generator sets. Other important manuals for this generator are also printed by this publisher.

An 8.5x11 3 hole punched loose leaf copy may be purchased for your 3 ring binder. Email books@ocotillopress.com for current information.

Should you have suggestions or feedback on ways to improve this book please send email to Books@OcotilloPress.com We also welcome requests for other military manuals you would like to see printed.

Edited 2021 Ocotillo Press
ISBN 978-1-954285-15-6

Printed in the United States of America

Ocotillo Press
Houston, TX 77017
Books@OcotilloPress.com

Disclaimer: The user of this book is responsible for following safe and lawful practices at all times. The publisher assumes no responsibility for the use of the content of this book. The publisher has made an effort to ensure that the text is complete and properly typeset, however omissions, errors, and other issues may exist that the publisher is unaware of.

*ARMY TM 9-6115-643-10
AIR FORCE TO 35C2-3-445-21

TECHNICAL MANUAL

OPERATOR'S MANUAL
FOR

GENERATOR SET, SKID MOUNTED, TACTICAL QUIET, 15 kW, 50/60 Hz, MEP-804A
(NSN: 6115-01-274-7388) (EIC: VG4)
15 kW, 50/60 Hz, MEP-804B
(NSN: 6115-01-530-1458) (EIC: N/A)

GENERATOR SET, SKID MOUNTED, TACTICAL QUIET, 15 kW, 400 Hz, MEP-814A
(NSN: 6115-01-274-7393) (EIC: VN4)
15 kW, 400 Hz, MEP-814B
(NSN: 6115-01-529-9494) (EIC: N/A)

*<u>SUPERSEDURENOTICE</u>- This manual supersedes TM 9-6115-643-10 dated 01 April 2008. Date of issue for the revised manual is: 15 February 2010.

HEADQUARTERS, DEPARTMENTS OF THE ARMY ANDAIRFORCE 15
FEBRUARY 2010

WARNING SUMMARY

FIRST AID

For First Aid information, refer to FM 4-25.11.

5   5 SAFETY STEPS TO FOLLOW IF SOMEONE IS THE VICTIM OF ELECTRICAL SHOCK:

1   DO NOT TRY TO PULL OR GRAB THE INDIVIDUAL.

2   IF POSSIBLE, TURN OFF THE ELECTRICAL POWER.

3   IF YOU CANNOT TURN OFF THE ELECTRICAL POWER, PULL, PUSH OR LIFT THE PERSON TO SAFETY USING A DRY WOODEN POLE OR A DRY ROPE OR SOME OTHER INSULATING MATERIAL.

4   SEND FOR HELP AS SOON AS POSSIBLE.

5   AFTER THE INJURED PERSON IS FREE OF CONTACT WITH THE SOURCE OF ELECTRICAL SHOCK, MOVE THE PERSON A SHORT DISTANCE AWAY AND IMMEDIATELY START ARTIFICIAL RESUSCITATION.

WARNING SUMMARY - Continued

WARNING AND CAUTION STATEMENTS

Warning and Caution statements have been strategically placed throughout this text prior to operating procedures, practices, or conditions considered essential to the protection of personnel (WARNING) or equipment and property (CAUTION).

A WARNING or CAUTION will apply each time the related step is repeated. Prior to starting any task the WARNINGs or CAUTIONs included in the text for that task must be reviewed and understood. Refer to the materials list at the beginning of the appropriate manual section for materials used during maintenance of this equipment. This warning summary contains the WARNINGs and CAUTIONs included in the manual. The detailed warnings for hazardous materials only are listed separately in the warning summary as the "Hazardous Materials Warnings" section.

WARNING

Metal jewelry can conduct electricity and become entangled in generator set components. Remove all metal jewelry when working on generator set. Failure to comply can cause injury or death to personnel.

WARNING

High voltage is produced when this generator set is in operation. Make sure unit is completely shut down and free of any power source before attempting any repair or maintenance on the unit. Failure to comply can cause injury or death to personnel.

WARNING

High voltage is produced when the generator set is in operation. Never attempt to start the generator set unless it is properly grounded. Failure to comply can cause injury or death to personnel.

WARNING

High voltage is produced when the generator set is in operation. Never attempt to connect or disconnect load cables while the generator set is running. Failure to comply can cause injury or death to personnel.

WARNING

Jumper will not be removed unless equipment being powered specifically required an isolated ground (floating ground). Failure to comply with this warning can cause injury or death to personnel.

WARNING

DC voltages are present at generator set electrical components even with generator set shut down. Avoid shorting any positive with ground/negative. Failure to comply can cause serious injury to personnel and damage to equipment.

WARNING

Slave receptacle (NATO connector) is electrically live at all times and is unfused. The Battery Disconnect Switch does not remove power from the slave receptacle. NATO slave receptacle has 24 VDC even when Battery Disconnect Switch is set to OFF. This circuit is only dead when the batteries are fully disconnected. Disconnect the batteries before performing maintenance on the slave receptacle. Failure to comply can cause serious injury or death to personnel.

WARNING SUMMARY - Continued
WARNING

High voltage power is available when the main contactor is closed. Avoid accidental contact with live components. Ensure load cables are properly connected and the load cable door is shut before closing main contactor. Ensure load is turned off before closing main contactor. Ensure that no one is working with/on loads connected to the generator set before closing main contactor. Failure to observe this warning can result in severe personal injury or death by electrocution.

WARNING

A qualified technician must make the power connections and perform all continuity checks. The power source may be a generator or commercial power. Failure to comply with this warning can result in injury or death to personnel.

WARNING

Do not wear loose clothing when performing checks, services and maintenance. Loose clothing may be entangled in generator set components. Failure to comply can cause injury or death to personnel.

WARNING

All fuel is combustible and toxic to eyes, skin, and respiratory tract. Skin and eye protection are required when working in contact with diesel fuel. Avoid repeated or prolonged contact. Provide adequate ventilation. Operators are to wash skin exposed to fuel and change fuel soaked clothing promptly. Failure to comply can cause serious injury to personnel.

WARNING

Fuels used in the generator set are combustible. When filling the fuel tank, maintain metal-to-metal contact between filler nozzle and fuel tank opening to eliminate static electrical discharge. Failure to comply can result in flames and possible explosion and can cause injury or death to personnel and damage to the generator set.

WARNING

Fuels used in the generator set are combustible. Do not smoke or use open flames when performing maintenance. Failure to comply can result in flames and possible explosion and can cause injury or death to personnel and damage to the generator set.

WARNING

Hot engine surfaces from the engine and generator circuitry are possible sources of ignition. When hot refueling with DF-1, DF-2, JP5 or JP8, avoid fuel splash and fuel spill. Do not smoke or use open flame when performing refueling. Remember PMCS is still required. Failure to comply can result in flames and possible explosion and can cause injury or death to personnel and damage to the generator set.

WARNING

If necessary to move a generator set that has been operating in parallel with another generator set, shut down all generator sets prior to removing load cables or ground. Failure to comply can cause injury or death to personnel by electrocution.

WARNING

Before making any connections for parallel operation or moving a generator set that has been operating in parallel, ensure there is no input to the load board and the generator sets are shut down. Failure to comply can cause injury or death to personnel by electrocution.

WARNING SUMMARY - Continued
WARNING

Cooling system operates at high temperature and pressure. Contact with high pressure steam and/or liquids can result in burns and scalding. Shut down generator set, and allow system to cool before performing checks, services and maintenance. Failure to comply can cause injury or death to personnel.

WARNING

Cooling system operates at high temperature and pressure. When performing DURING PMCS, wear gloves and additional protective clothing and goggles as required. Contact with high pressure steam and/or liquids can result in burns and scalding.

WARNING

In extreme cold weather, skin can stick to metal. Avoid contacting metal items with bare skin in extreme cold weather. Failure to comply can cause injury or death to personnel.

WARNING

Operating the generator set exposes personnel to a high noise level. Hearing protection must be worn when operating or working near the generator set when the generator set is running. Failure to comply can cause hearing damage to personnel.

WARNING

When running, generator set engine has hot metal surfaces that will burn flesh on contact. Shut down generator set, and allow engine to cool before checks, services and maintenance. Wear gloves and additional protective clothing as required. Failure to comply can cause serious injury to personnel.

WARNING

When running, generator set engine has hot metal surfaces that will burn flesh on contact. When performing DURING PMCS, wear gloves and additional protective clothing as required. Failure to comply can cause injury or death to personnel.

WARNING

Exhaust discharge contains deadly gases including carbon monoxide. Do not operate generator set in an enclosed area unless exhaust discharge is properly vented outside. Failure to comply can cause injury or death to personnel.

WARNING

Hot exhaust gases can ignite flammable materials. Allow room for safe discharge of hot gases and sparks. Failure to comply can cause injury or death to personnel.

WARNING

Exhaust system can get very hot. Shut down generator set, and allow system to cool before performing checks, services and maintenance. Failure to comply can cause severe burns and injury to personnel.

WARNING

Exhaust system can get very hot. When performing DURING PMCS, wear gloves and additional protective clothing as required. Failure to comply can cause severe burns and injury to personnel.

WARNING SUMMARY - Continued
WARNING

When running, winterization heater has hot metal surfaces that will burn flesh on contact. Shut down generator set and allow heater to cool before performing maintenance. Wear gloves and additional protective clothing as required. Failure to comply can cause serious injury to personnel.

WARNING

Top housing panels can get very hot. Allow panels to cool down before performing maintenance. Failure to comply can result in severe burns to personnel.

WARNING

Top housing panels can get very hot. When performing DURING PMCS, wear gloves and additional protective clothing as required. Failure to comply can result in severe burns to personnel.

WARNING

Eye protection is required when working with compressed air. Compressed air can propel particles at high velocity and injure eyes. Do not exceed 15 psi pressure when using compressed air. Failure to comply could cause serious injury to personnel.

WARNING

Cleaning compound is toxic. Avoid prolonged breathing of vapors. Use only in a well-ventilated area. Failure to comply can cause serious injury to personnel.

WARNING

When disconnecting or removing batteries, disconnect the negative lead that connects directly to the grounding stud first. Disconnect the negative end of the interconnection cable next. When installing batteries, reverse the connection sequence. Failure to comply can cause serious injury to personnel.

WARNING

Batteries give off a flammable gas. Do not smoke or use open flame when performing maintenance. Failure to comply can cause injury or death to personnel and equipment damage due to flames and explosion.

WARNING

Battery acid can cause burns to skin and cause eye injury. Wear safety goggles and chemical gloves and avoid acid splash while working on the batteries. Failure to comply may cause injury or death to personnel.

WARNING

Do not attempt to lift, carry, or move the generator set, assemblies or large components yourself. Observe the decals on equipment which identify the weight and determine if an assistant is needed. Maximum weight for one person is no more than 16.81 kilograms (kg) (37 pounds (lb)). Failure to comply may cause serious injury to personnel.

WARNING

Do not operate generator set if fuel leaks are present. Fuel is combustible. Always perform PMCS before operation. Failure to comply may cause fire or explosion and injury or death to personnel.

WARNING SUMMARY - Continued
WARNING

Generator set operator is permitted to make connections only to output terminal board. Connections to load distribution points or equipment beyond the output box shall only be made by properly trained and authorized personnel. Failure to comply may cause injury or death to personnel.

WARNING

Generator set must be completely shut down prior to washing. Failure to comply may cause injury or death to personnel.

WARNING

Do not start generator set until internal components are completely dry. Failure to comply may cause injury or death to personnel.

WARNING

Power is available to the convenience receptacle when the generator set is running. Avoid accidental contact. Failure to comply may cause injury or death to personnel.

WARNING

Ensure the frequency of any device powered by the convenience receptacle matches the frequency of the generator set. Failure to comply may cause serious injury to personnel.

WARNING

Wear heat resistant gloves and avoid contacting hot metal surfaces with hands and exposed skin after components have been heated. Wear additional protective clothing as required. Failure to comply may cause injury or death to personnel.

LIST OF EFFECTIVE PAGES / WORK PACKAGES

NOTE: This manual supersedes TM 9-6115-643-10 dated 01 April 2008. Date of issue for the revised manual is: 15 February 2010. Zero in the "Change No." column indicates an original page or work package.

Date of issue for revision is:

Original 15 FEBRUARY 2010

TOTAL NUMBER OF PAGES FOR FRONT AND REAR MATTER IS 34 AND TOTAL NUMBER OF WORK PACKAGES IS 28, CONSISTING OF THE FOLLOWING:

Page / WP No. Change No.		Page / WP No.	Change No.
Front Cover	0	WP 0026 (4 pgs)	0
Blank	0	WP 0027 (3 pgs)	0
Warning summary (6 pgs)	0	WP 0028 (1 pgs)	0
i-x	0	INDEX-1 - INDEX-2	0
Chp 1 title page	0	Inside back cover	0
Index	0	Back cover	0
WP 0001 (4 pgs)	0		
WP 0002 (9 pgs)	0		
WP 0003 (10 pgs)	0		
Chp 2 title page	0		
Index	0		
WP 0004 (6 pgs)	0		
WP 0005 (22 pgs)	0		
WP 0006 (5 pgs)	0		
WP 0007 (2 pgs)	0		
Chp 3 title page	0		
Index	0		
WP 0008 (2 pgs)	0		
WP 0009 (9 pgs)	0		
Chp 4 title page	0		
Index	0		
WP 0010 (4 pgs)	0		
WP 0011 (8 pgs)	0		
WP 0012 (2 pgs)	0		
WP 0013 (2 pgs)	0		
WP 0014 (2 pgs)	0		
WP 0015 (2 pgs)	0		
WP 0016 (2 pgs)	0		
WP 0017 (1 pgs)	0		
WP 0018 (2 pgs)	0		
Chp 5 title page	0		
Index	0		
WP 0019 (2 pgs)	0		
WP 0020 (5 pgs)	0		
WP 0021 (1 pgs)	0		
WP 0022 (2 pgs)	0		
Chp 6 title page	0		
Index	0		
WP 0023 (1 pgs)	0		
WP 0024 (3 pgs)	0		
WP 0025 (2 pgs)	0		

HEADQUARTERS,
DEPARTMENTS OF THE ARMY
AND AIR FORCE
WASHINGTON, D.C., 15 FEBRUARY 2010

TECHNICAL MANUAL

OPERATOR'S MANUAL

GENERATOR SET, SKID MOUNTED, TACTICAL QUIET,
15 kW, 50/60 Hz, MEP-804A
(NSN: 6115-01-274-7388) (EIC: N/A)
15 kW, 50/60 Hz, MEP-804B
(NSN: 6115-01-530-1458) (EIC: N/A)

GENERATOR SET, SKID MOUNTED, TACTICAL QUIET,
15 kW, 400 Hz, MEP-814A
(NSN: 6115-01-274-7393) (EIC: N/A)
15 kW, 400 Hz, MEP-814B
(NSN: 6115-01-529-9494) (EIC: N/A)

REPORTING ERRORS AND RECOMMENDING IMPROVEMENTS

You can help improve this manual. If you find any mistakes or if you know of a way to improve the procedures, please let us know. Reports, as applicable by the requiring Service, should be submitted as follows:

(a)(A) Army - Mail your letter or DA Form 2028 (Recommended Changes to Publications and Blank Forms), located in the back of this manual, directly to: Commander, U.S. Army CECOM (LCMC) and Fort Monmouth, ATTN: AMSEL-LC-LEO-E-CM, Fort Monmouth, NJ 07703-5006. You may also send in your recommended changes via electronic mail or by fax. Our fax number is 732-532-3421, DSN 992-3421. Our e-mail address is MONM-AMSELLEOPUBSCHG@conus.army.mil. Our online web address for entering and submitting DA Form 2028s is http://edm.monmouth.army.mil/pubs/2028.html.

---

*SUPERSEDURENOTICE- This manual supersedes TM 9-6115-643-10 dated 01 April 2008. Date of issue for the revised manual is: 15 February 2010.

TABLE OF CONTENTS

TABLE OF CONTENTS - Continued

TABLE OF CONTENTS - Continued

TABLE OF CONTENTS - Continued

TABLE OF CONTENTS - Continued

How to Use This Manual

This manual contains operator maintenance instructions for the MEP-804A, MEP-804B, MEP-814A and MEP-814B Skid Mounted, Tactical Quiet Generator (TQG) Sets.

NOTE

Throughout the family of manuals, directional orientation in relation to the equipment is described from the point of view of the operator facing the operator's controls looking out over the equipment. From this perspective, the end of the equipment containing the operator's controls will be referred to as the rear.

This manual provides operating procedures, troubleshooting, maintenance, and supporting information required to operate and maintain the Skid Mounted, Tactical Quiet Generator Sets. Listed below are some of the features included in this TM to help locate and use the provided information.

WORK PACKAGES

This TM has been organized using the WP format. Each chapter contains a series of WPs rather than sections and paragraphs. Each WP is designed to stand alone as a complete information module; if the user keeps the section(s) of this TM in a loose-leaf binder, the user will be able to remove just the WP needed to complete a specific task. Here are some WP features of which the user should be aware.

Each WP is numbered using a four-digit number beginning with WP 0001. WPs are numbered sequentially throughout the TM (ex. WP 0016. WP 0020. etc.). The Table of Contents lists each chapter and WP title as well as all figures and tables contained within each. Figures and tables are numbered sequentially for each WP.

The WP number is located at the top right of each page. It is also located at the bottom of the page with the WP page number included (0001-1 would be page 1 of the General Information WP (WP 0001, General Information)).

Each WP starts on a right-hand page. This is done so the user can remove a single WP from the paper TM if needed for a task. Blank pages are assigned a number, but it appears on the preceding or following page. For example. if page 0001-10 of a WP is blank. page 0001-9 will have the number 0001-9/10 blank; or if page 0001-1 of a WP is blank, page 0001-2 will have the number 0001-1 blank/2.

Each WP containing step-by-step maintenance or troubleshooting procedures will end with the words END OF TASK, and each WP ends with the statement END OF WORK PACKAGE. Think of each WP as a small, stand-alone TM.

Typographical conventions are as follows:

[Unload] indicates a soft key or a switch.

[Previous] + [Next] indicates two simultaneous key presses. [+] [-] indicates two sequential key presses.

References to equipment Data and Description Plates are printed as they appear on the equipment whenever possible.

Warnings, Cautions and Notes Definitions

Warnings, cautions, notes. chapter titles, and paragraph headings are printed in bold type. Icons related to warnings are shown directly above the warning text.

The following definitions apply to WARNINGS, CAUTIONS and NOTES found throughout this publication. Warning, cautions and notes provide supplemental information. Personnel must understand and apply these Warnings, Cautions and Notes during many phases of operation and maintenance to ensure personnel safety and health and the protection of property. Portions of this information may be repeated in certain chapters of this publication for emphasis.

WARNING

A warning identifies a clear danger to the person doing that procedure.

CAUTION

A caution identifies risk of damage to the equipment.

NOTE

A note highlights essential procedures, conditions, or statements or conveys important instructional data to the user.

CHAPTER OVERVIEW

Chapter 1 - General Information, Equipment Description and Theory of Operation

Chapter 1 provides an introduction to the Skid Mounted, Tactical Quiet Generator Sets. It is divided into three work packages, as follows:

General Information. This work package provides general information about this manual and the related forms and records. Instructions are provided for making equipment improvement recommendations. Coverage includes a reference to the TM that contains instructions on destruction of materiel to prevent enemy use. Also, a list of abbreviations and acronyms is provided. Also, a nomenclature cross-reference list is provided as well as a list of abbreviations and acronyms.

Equipment Description and Data. This work package describes capabilities, characteristics, and features. It provides basic equipment data and shows the locations of major components. Descriptions of the major components are also provided.

Theory of Operation. This work package provides functional descriptions of the equipment.

Chapter 2 - Operator Instructions

Chapter 2 provides instructions for operating the Skid Mounted, Tactical Quiet Generator Sets. The chapter is divided into three work packages, as follows:

Description and Use of Operator Controls and Indicators. This work package provides references to the applicable generator set technical manuals and trailer technical manuals. Those references contain information on operator's controls and indicators for the equipment.

Operation Under Usual Conditions. This work package contains instructions for preparing the equipment for use and operation under normal conditions. Coverage includes connection instructions and preparation instructions for movement to a new worksite.

Operation Under Unusual Conditions. This work package provides unusual operating procedures or references to the applicable accompanying technical manuals.

Chapter 3 - Operator Troubleshooting Procedures

Chapter 3 covers troubleshooting procedures of the Skid Mounted, Tactical Quiet Generator Sets to be performed by the operator. The chapter is divided as follows:

Operator Troubleshooting Index. This work package provides a troubleshooting introduction and malfunction/symptom index to direct you to the appropriate troubleshooting procedure at the operator level.

Operator Troubleshooting Procedures. This work package provides troubleshooting procedures and corrective actions that are to be performed by the operator. It also provides references to the applicable technical manuals.

Chapter 4 - Operator Maintenance Instructions

Chapter 4 covers maintenance procedures for the Skid Mounted, Tactical Quiet Generator Sets to be performed by the operator. Its purpose is to provide you with the information that you need to keep the equipment in good operating condition. The chapter is divided as follows:

Operator Preventive Maintenance Checks and Services (PMCS) Introduction. This work package provides a detailed explanation of each table entry in the PMCS table along with applicable warnings, cautions and notes prior to starting on the PMCS procedures.

Operator Preventive Maintenance Checks and Services (PMCS). This work package contains detailed instructions that the operator must perform before, during, and after preventive maintenance checks and services. Coverage includes all operator PMCS for the equipment.

Operator Lubrication Instructions. This work package section provides references to the applicable lubrication instructions.

Operator Maintenance Procedures. These work packages refer the operator to the preventive maintenance checks and services required by WP 0011.

Chapter 5 - Supporting Information

Chapter 4 covers maintenance procedures for the Skid Mounted, Tactical Quiet Generator Sets to be performed by the operator. Its purpose is to provide you with the information that you need to keep the equipment in good operating condition. The chapter is divided as follows:

Components of End Item (COEI) and Basic Issue Items (BII) Lists. This work package lists the items usually packaged separately but needed for installation and operation of the equipment. The work package has three sections, as follows:

> **Introduction.** This section explains the entries in Tables 1 and 2.

> **Components of End Item.** The equipment is normally shipped fully assembled, so this section is not applicable.

> **Basic Issue Items.** This section contains a list of the accessories needed for installation and operation of the equipment.

Additional Authorization List (AAL). This work package lists additional items you are authorized for support of the equipment. This work package contains two sections, as follows:

> **Introduction.** This section explains the entries in Tables 1.

> **Additional Authorized Items List.** This table lists the Additional Authorized Items.

Expendable and Durable Items List. This work package lists expendable/durable supplies and materials needed to operate and maintain your equipment. The work package contains two sections, as follows:

> **Introduction.** This section explains the entries in Tables 1.

> **Expendable and Durable Items List.** The list indicates the maintenance level that needs each item and identifies the items by National Stock Number (NSN), description, and unit of measure.

Definition of Unusual Terms. This section lists and defines the terms used in this technical manual that are not listed in the Army Regulation (AR 310-25).

Rear Matter

Alphabetical Index. An alphabetical index at the back of this technical manual provides a listing of subjects covered, cross-referenced to the applicable work packages.

HOW TO FIX AN EQUIPMENT MALFUNCTION

Determining the Cause

Finding the cause of a malfunction, troubleshooting, is the first step in fixing your equipment and returning it to operation. Follow these simple steps to determine the root of the problem:

1. Turn to the Table of Contents in this manual.
2. Locate "Troubleshooting" under the chapter that covers your level of maintenance. Turn to the page indicated.
3. For operator troubleshooting, find the malfunction listing in the troubleshooting symptom index. Follow the instructions provided as indicated by the symptom index.

Preparing for a Task

Be sure that you understand the entire maintenance procedure before beginning any maintenance task. Make sure that all parts, materials, and tools are handy. Read all steps before beginning.

Prepare to do the task as follows:

1. Carefully read the entire task before starting. It tells you what you will need and what you have to know to start the task. DO NOT START THE TASK UNTIL:
   a. You know what is needed
   b. You have everything you need

c. You understand what to do

2. If parts are listed, they can be drawn from technical supply. Before you start the task, check to make sure you can get the needed parts.

3. If expendable/durable supplies or materials are needed, get them before starting the task. Refer to WP 0022 for the correct nomenclature and NSN.

How to Do the Task

Before starting, read the entire task. Be sure that you understand the entire procedure before you begin the task. As you read, remember the following:

1. PAY ATTENTION TO WARNINGS, CAUTIONS, AND NOTES.

2. Use the List of Abbreviations/Acronyms if you do not understand the special abbreviations or unusual terms used in this manual.

3. The following are standard maintenance practices. Instructions about these practices are usually not included in task steps. When standard maintenance practices do not apply, the task steps will tell you.

   a. Discard used preformed packing, retainers, gaskets, cotter pins, lockwashers, and similar items. Install new parts to replace the discarded items.

   b. Coat packing before installation, in accordance with the task instructions.

   c. Disassembly procedures describe the disassembly needed for total authorized repair. You may not need to disassemble an item as far as described in the task. Follow the disassembly steps only as far as needed to repair/replace worn or damaged parts.

   d. Clean the assembly, subassembly, or part before inspecting it.

   e. Before installing components having mating surfaces, inspect the mating surfaces to make sure they are in serviceable condition.

   f. Hold the bolt (or screw) head with a wrench (or screwdriver) while tightening or loosening a nut on the bolt (or screw).

   g. When a cotter pin is required, align the cotter pin holes within the allowable torque range.

   h. Inspect for foreign objects after performing maintenance.

CHAPTER 1

OPERATOR GENERAL INFORMATION, EQUIPMENT DESCRIPTION AND THEORY OF OPERATION

FOR

15 kW 50/60 AND 400 Hz SKID MOUNTED, TACTICAL QUIET GENERATOR SET

CHAPTER 1

GENERAL INFORMATION, EQUIPMENT DESCRIPTION AND THEORY OF OPERATION

WORK PACKAGE INDEX

OPERATOR MAINTENANCE

15 kW 50/60 AND 400 Hz SKID MOUNTED, TACTICAL QUIET GENERATOR SET

GENERAL INFORMATION

SCOPE

Type of Manual

This manual contains operation and operator maintenance instructions for the Tactical Quiet (TQ), 15 kW 50/60 and 400 Hz Generator Sets (Figure 1), herein referred to as generator set. Included are descriptions of major components and their functions in relation to other components.

Purpose of Equipment

The generator set provides tactical quiet AC power. The generator set is easily transported, operated, and maintained.

MAINTENANCE FORMS, RECORDS, AND REPORTS

(1) (A) Department of the Army forms and procedures used for equipment maintenance will be those prescribed by (as applicable) DA PAM 750-8, The Army Maintenance Management System (TAMMS) Users Manual; DA PAM 738-751, Functional Users Manual for the Army Maintenance Management Systems - Aviation (TAMMS-A); or AR 700-138, Army Logistics Readiness and Sustainability.

(2) (F) Maintenance forms and records used by Air Force personnel are prescribed in AFI 21-101 and the applicable TO 00-20 Series Technical Orders.

(3) (N) Navy users should refer to their service peculiar directives to determine applicable maintenance forms and records to be used.

REPORTING EQUIPMENT IMPROVEMENT RECOMMENDATION (EIR)

If your Generator Set needs improvement, let us know. Send us an EIR. You, the user, are the only one who can tell us what you don't like about your equipment. Let us know why you don't like the design or performance. If you have Internet access, the easiest and fastest way to report problems or suggestions is to go to https://aeps.ria.army.mil/aepspublic.cfm (scroll down and choose the "Submit Quality Deficiency Report" bar). The Internet form lets you choose to submit an Equipment Improvement Recommendation (EIR), a Product Quality Deficiency Report (PQDR) or a Warranty Claim Action (WCA). You may also submit your information using an SF 368 (Product Quality Deficiency Report). You can send your SF 368 via e-mail, regular mail, or facsimile using the addresses/facsimile numbers specified in DA PAM 750-8, The Army Maintenance Management System (TAMMS) Users Manual. We will send you a reply.

CORROSION PREVENTION AND CONTROL (CPC)

Corrosion Prevention and Control (CPC) of Army materiel is a continuing concern. It is important that any corrosion problems with this item be reported so that the problem can be corrected and improvements can be made to prevent the problem in future items.

Corrosion specifically occurs with metals. It is an electrochemical process that causes the degradation of metals. It is commonly caused by exposure to moisture, acids, bases, or salts. An example is the rusting of iron. Corrosion damage in metals can be seen, depending on the metal, as tarnishing, pitting, fogging, surface residue, and/or cracking.

Plastics, composites, and rubbers can also degrade. Degradation is caused by thermal (heat), oxidation (oxygen), solvation (solvents), or photolytic (light, typically UV) processes. The most common exposures are excessive heat or light. Damage from these processes will appear as cracking, softening, swelling, and/or breaking.

SF Form 368, Product Quality Deficiency Report should be submitted to the address specified in DA PAM 750-8, The Army Maintenance Management System (TAMMS) Users Manual.

DESTRUCTION OF ARMY MATERIEL TO PREVENT ENEMY USE

Destruction of Army materiel to prevent enemy use shall be in accordance with TM 750-244-3.

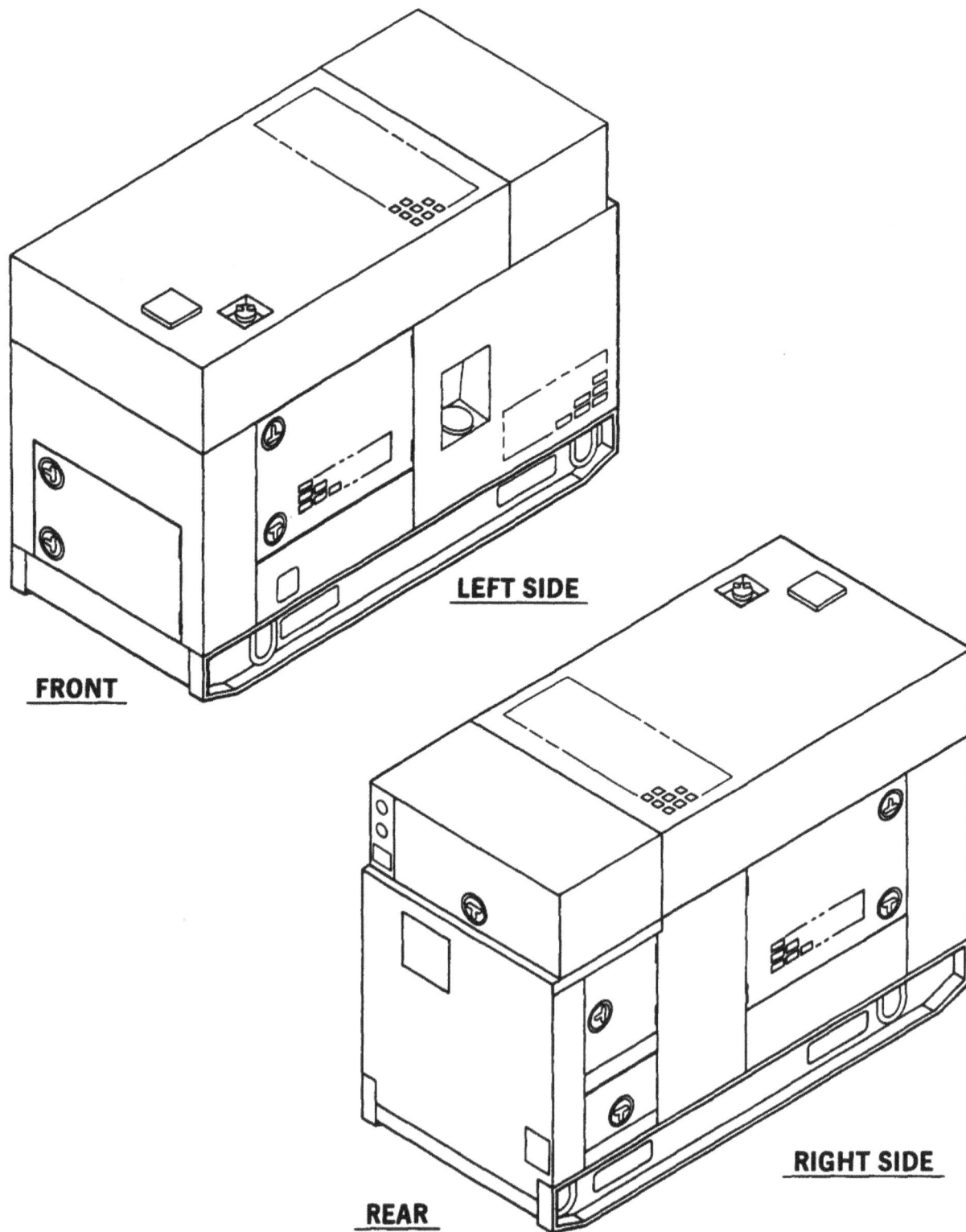

LEFT SIDE

FRONT

REAR

RIGHT SIDE

Figure 1. Generator Set, 15 kW, Tactical Quiet.

PREPARATION FOR STORAGE OR SHIPMENT

Information on Preparation for Storage or Shipment, refer to WP 0005, Preparation for Movement.

WARRANTY INFORMATION

Generator sets MEP-804A/MEP-814A manufactured under contract number DAAK01-88-D-D082 are warranted by Libby Corporation for a period of 36 months or 1800 operating hours, whichever occurs first. Generator sets MEP-804A/MEP-814A manufactured under contract number DAAK01-94-D-0036 and MEP-804B/MEP-814B manufactured under contract number DAAK01-97-D-0034 are warranted by Fermont, Inc. for a period of 36 months or 1800 operating hours, whichever occurs first. Refer to Warranty Technical Bulletin TB 9-6115-643-24. The warranty starts on the date found in block 23, DA Form 2408-9, in the logbook. Report all defects in material or workmanship to your supervisor, who will take appropriate action through your Unit Maintenance Shop.

NOMENCLATURE CROSS-REFERENCE LIST

CommonName OfficialNomenclature

MEP-804A/MEP-804B Generator Set, Skid Mounted, Tactical Quiet, 15 kW 50/60 Hz

MEP-814A/MEP-814B Generator Set, Skid Mounted, Tactical Quiet, 15 kW 400 Hz

LIST OF ABBREVIATIONS/ACRONYMS

Abbreviation/Acronym Name

° C   Degrees Celsius

° F   Degrees Fahrenheit AAL Additional Authorization List

AOAP Army Oil Analysis Program

BII Basic Issue Item

BOI Basis Of Issue

CAGE Commercial And Government Entity

CAGEC Commercial And Government Entity Code

COEI Components Of End Item

CPC Corrosion Prevention and Control

CTA Common Table Of Allowance

DMWR Depot Maintenance Work Requirement

DOD Department Of Defense

EIR Equipment Improvement Recommendation

FGC Functional Group Code ft•lbf Foot-Pound Force

Hz   Hertz

JTA Joint Table Of Allowances kg   Kilogram kPa

Kilopascals kVA   Kilovolt-ampere kW Kilowatt

LIST OF ABBREVIATIONS/ACRONYMS - Continued

Abbreviation/Acronym Name

m Meter (Metric Measure)

MTOE Modified Table Of Organization and Equipment

NATO North Atlantic Treaty Organization

NHA Next Higher Assembly

NIIN National Item Identification Number

NSN National Stock Number N•m Newton Meter P/N

Part Number

PMCS Preventive Maintenance Checks and Services

SMR Source, Maintenance, and Recoverability

TAMMS The Army Maintenance Management System

UOC Usable On Code

END OF WORK PACKAGE

OPERATOR MAINTENANCE

15 kW 50/60 AND 400 Hz SKID MOUNTED, TACTICAL QUIET GENERATOR SET

EQUIPMENT DESCRIPTION AND DATA

EQUIPMENT CHARACTERISTICS, CAPABILITIES, AND FEATURES

The generator sets, MEP-804A/MEP-814A (Figure 1) and MEP-804B/MEP-814B (Figure 2), are fully enclosed, self-contained, skid-mounted, portable units. They are equipped with controls, instruments and accessories necessary for operation as single units or in parallel with another unit of the same class and mode. The generator sets consist of a diesel engine, brushless generator, excitation system, speed governing system, fuel system, 24 VDC starting system, control system and fault system.

LOCATION AND DESCRIPTION OF MAJOR COMPONENTS

NOTE

All locations (index numbers) referenced are given facing the control panel assembly (rear) of the generator set. These index numbers refer to both Figure 1 (MEP-804A/MEP-814A) and Figure 2 (MEP-804B/MEP-814B), except for the turbocharger (Figure 2).

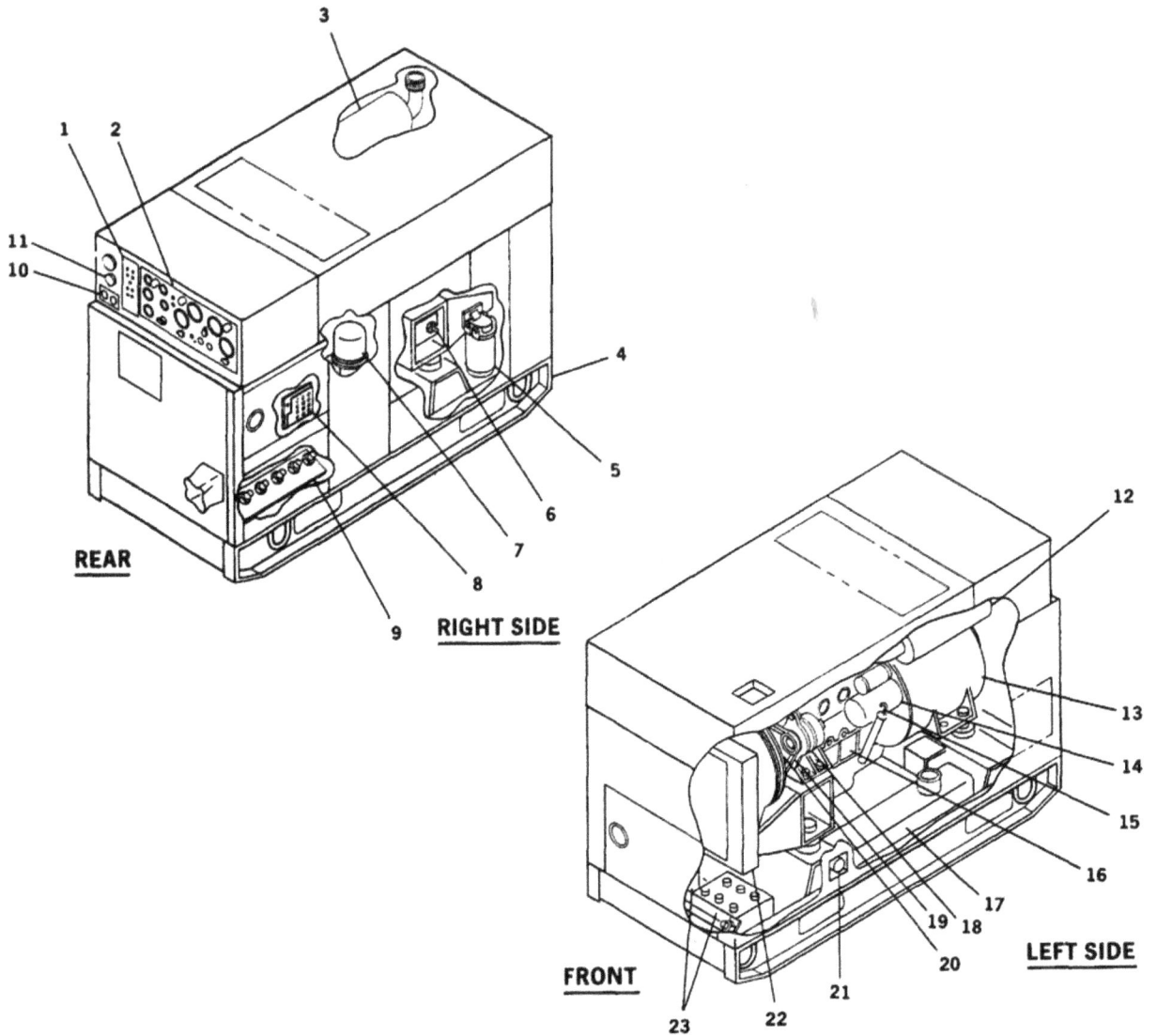

LEGEND

1 Malfunction Indicator Panel 13 AC Generator
2 Control Panel Assembly 14 Starter
3 Muffler 15 Dipstick
4 Skid Base 16 Engine
5 Fuel filter/Water Separator 17 Fuel Tank
6 Dead Crank Switch 18 Battery charging Alternator
7 Oil Filter 19 Fan Belt
8 Voltage Reconnection Terminal Board 20 Water Pump
9 Load Output Terminal Board 21 NATO Slave Receptacle
10 Convenience Receptacle 22 Radiator
11 Paralleling Receptacle 23 Batteries
12 Air Cleaner Assembly

Figure 1. Generator Set Components - MEP-804A/MEP-814A.

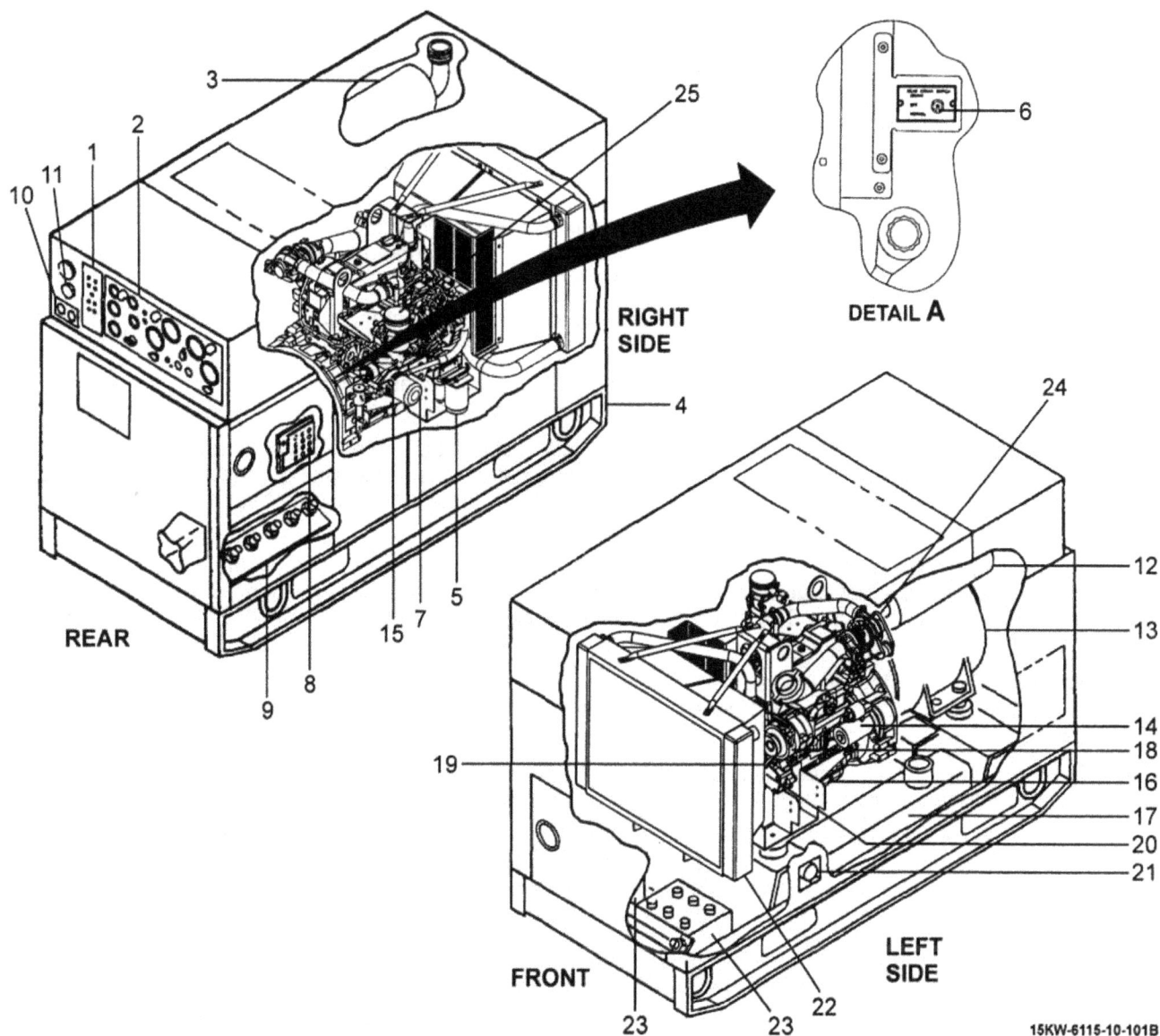

DETAIL A

RIGHT SIDE

REAR

FRONT

LEFT SIDE

15KW-6115-10-101B

LEGEND
1 Malfunction Indicator Panel 14 Starter
2 Control Panel Assembly 15 Dipstick
3 Muffler 16 Engine
4 Skid Base 17 Fuel Tank
5 Fuel filter/Water Separator 18 Battery charging Alternator
6 Dead Crank Switch 19 Fan Belt
7 Oil Filter 20 Water Pump
8 Voltage Reconnection Terminal Board 21 NATO Slave Receptacle
9 Load Output Terminal Board 22 Radiator
10 Convenience Receptacle 23 Batteries
11 Paralleling Receptacle 24 Turbocharger
12 Air Cleaner Assembly 25 Crankcase Ventilation Filter
13 AC Generator

Figure 2. Generator Set Components - MEP-804B/MEP-814B.

Engine(16). The generator is powered by one of two possible engines: an Isuzu engine (MEP- 804A/MEP-814A) or a Yanmar engine (MEP-804B/MEP-814B). The engine occupies the front half of the generator set. The Isuzu engine is a four cylinder, four cycle, fuel injected, naturally aspirated, liquid-cooled diesel engine. The Yanmar engine is similar, but turbo-charged. The engine is also equipped with a fuel filter/water separator, oil filter, and an air cleaner assembly. Protection devices automatically stop the engine during conditions of high coolant temperature, low oil pressure, no fuel, over-speed, or over-voltage.

Radiator(22). The radiator is located at the front of the generator set. It acts as a heat exchange for the engine coolant.

Muffler(3). On the Isuzu engine, the muffler and exhaust tubing are connected to the exhaust manifold on the engine. On the Yanmar engine, the muffler and exhaust tubing are connected to the turbocharger. The exhaust exits from the top of the generator set housing. Gases are exhausted upward.

Starter(14). The starter is located on the left side of the engine. The electric starter mechanically engages and the engine flywheel in order to start the diesel engine.

BatteryChargingAlternator(18). The battery charging alternator is located on the left side of the engine. It is capable of maintaining the batteries in a state of full charge in addition to providing the required 24 VDC control power.

Batteries(23). Two batteries are located at front of the generator set. The batteries are electrolyte serviceable, lead acid, 12 volt type, connected in series. After starting, the generator set is capable of operating with batteries removed. A diode and a fuse, located behind the control panel assembly, protect the generator set if the batteries are incorrectly connected.

AirCleanerAssembly(12). The air cleaner assembly is located on the left side behind the air cleaner access door. It consists of a dry-type, disposable paper element and canister. The air cleaner assembly features a dust collector which traps large dust particles. The air cleaner assembly has a restriction indicator which will indicate when the air cleaner element requires servicing.

FuelTank(17). The fuel tank is located below the engine and between the skid base side members. The fuel tank has a capacity of 14 gallons (53 liters) which will allow the generator set to operate for at least 8 hours without refueling.

ACGenerator(13). The AC generator is a single bearing, synchronous, brushless, three phase, fancooled generator. The generator is coupled directly to the rear of the diesel engine.

LoadOutputTerminalBoard(9). The load output terminal board is located on the right side (rear) of the generator set. There are four output terminals located on the board. They are marked L1, L2, L3 and LO. A fifth terminal, marked GND, is located next to the output terminals and serves as equipment ground for the generator set. A removable, solid copper bar is connected between the LO and GND terminals.

ControlPanelAssembly(2). The generator set control panel assembly is located at the rear of the generator set and contains controls and instruments for operating the engine and the generator.

MalfunctionIndicatorPanel(1). The malfunction indicator panel is located to the left of the control panel assembly. It indicates malfunctions of the generator set components.

NATOSlaveReceptacle(21). The NATO slave receptacle is located on the left side of the generator set under engine compartment access door. It is used for slave starting.

SkidBase(4). The skid base supports the generator set. It has fork lift access openings and cross members for short distance movement. The skid base has provisions in the bottom for installation of the generator set on a trailer.

VoltageReconnectionTerminalBoard(8). The voltage reconnection terminal board is located on the right side (rear) of the generator set. The board allows reconfiguration from 120/208 to 240/416 VAC output.

FuelFilter/WaterSeparator(5). The fuel filter/water separator is located in the engine compartment on the right side. The element removes impurities and water from the diesel fuel.

Dipstick(15). On the Isuzu engine, the dipstick is located on the left side of the engine compartment. On the Yanmar engine, the dipstick is located on the right side of the engine compartment. The dipstick shows the lubricating oil level in the engine crankcase.

OilFilter(7). The oil filter is located in the engine compartment on the right side. The filter removes impurities from the engine lubricating oil.

FanBelt(19). The fan belt is located in the engine compartment on the front of the engine. The belt drives the fan, water pump and battery charging alternator.

WaterPump(20). The water pump is located in the engine compartment on the front of the engine. The pump circulates the engine coolant through the engine block and the radiator.

DeadCrankSwitch(6). The Dead Crank switch is located in the engine compartment on the right side. The switch allows the engine to be cranked without starting for maintenance purposes.

ParallelingReceptacle(11). The Paralleling receptacle is used to connect the paralleling cable between two generator sets of the same size and mode to operate in parallel.

ConvenienceReceptacle(10). The convenience receptacle is a 10 Amp, 120 VAC receptacle used to operate small plug in type equipment. It is protected by a Ground Fault Circuit Interrupter located below the Malfunction Indicator Panel (Malfunction Indicator Panel (1)), an Overload Circuit Breaker located inside the control box, and an inline fuse on generator sets, contract number DAAKO1-88-D-0082. The convenience receptacle power is available at all times during operation of the generator set.

Turbocharger(24). The Yanmar engine in the MEP-804B/MEP-814B generator sets is turbocharged. The turbocharger increases the horsepower of the diesel engine in order to deliver the generator set maximum power.

CrankcaseVentilationFilter(25). The Yanmar engine in the MEP-804B/MEP-814B generator sets contains a crankcase ventilation filter located on the right side of the engine. The filter separates oil mist from the crankcase gases. The clean gases are returned to the intake manifold and subsequently burned. The collected oil is returned to the crankcase. This process reduces emissions and engine oil consumption.

WinterizationKit. See Chapter 6 for detailed information and breakdown.

DIFFERENCES BETWEEN MODELS

The differences between models of the generator sets covered in this manual are as follows: Model MEP-804A is equipped with a 50/60 Hz generator and uses an Isuzu diesel engine.
Model MEP-804B is equipped with a 50/60 Hz generator and uses a Yanmar turbocharged diesel engine.
Model MEP-814A is equipped with a 400 Hz generator and uses an Isuzu diesel engine.
Model MEP-814B is equipped with a 400 Hz generator and uses a Yanmar turbocharged diesel engine.

EQUIPMENT DATA

For a list of Leading Particulars refer to Table 1.

Table 1. Leading Particulars.

1. Generator Set:	
Model Number	
15 kW 50/60 Hz with Isuzu Engine	MEP-804A
15 kW 50/60 Hz with Yanmar Engine	MEP-804B
15 kW 400 Hz with Isuzu Engine	MEP-814A
15 kW 400 Hz with Yanmar Engine	MEP-814B
National Stock Number	
MEP-804A	6115-01-274-7388
MEP-804B	6115-01-530-1458
MEP-814A	6115-01-274-7393
MEP-814B	6115-01-529-9494
Overall Length	
MEP-804A/MEP-804B	69.7 in. (177.2 cm.)

Table 1. Leading Particulars. - Continued

MEP-814A/MEP-814B	69.7 in. (177.2 cm.)
Overall Width	
MEP-804A/MEP-804B	35.7 in. (90.8 cm.)
MEP-814A/MEP-814B	35.7 in. (90.8 cm.)
Overall Height	
MEP-804A/MEP-804B	55 in. (139.7 cm.)
MEP-814A/MEP-814B	55 in. (139.7 cm.)
Overall Weights (less Basic Issue Items)	
MEP-804A	1885 lb. (855.0 kg.)
MEP-804B	1785 lb. (809.6 kg.)
MEP-814A	2015 lb. (914.0 kg.)
MEP-814B	1915 lb. (868.6 kg.)
Wet Weights	
MEP-804A	2140 lb. (970.7 kg.)
MEP-804B	2040 lb. (925.3 kg.)
MEP-814A	2250 lb. (1020.6 kg.)
MEP-814B	2150 lb. (975.2 kg.)
2. Engine (MEP-804A/MEP-814A):	
Manufacturer	Isuzu
Model	C-240
Type	Four cylinder, four cycle, naturally aspirated diesel
Displacement	145 cu. in. (2.4 liters)
Altitude Degradation, 4000 ft. (1220 m.) to 8000 ft. (2440 m.)	3.5% per 1000 ft. (305 m.)
Firing Order	1, 3, 4, 2
Cold Weather Starting Aid System Use	40°F(4°C) or below
Valve Tappet Clearance Adjustment	
Hot or Cold (Intake)	0.045 in. (12 mm.)
Hot or Cold (Exhaust)	0.018 in. (0.45 cm.)
3. Engine (MEP-804B/MEP-814B):	
Manufacturer	Yanmar
Model	4TNV84T
Part Number	4TNV84T-DFM
Type	Four cylinder, four cycle, turbocharged diesel
Displacement	121.7 cu. in. (1.995 liters)
Altitude Degradation, 4000 ft. (1220 m.) to 8000 ft. (2440 m.)	3.5% per 1000 ft. (305 m.)
Firing Order	1, 3, 4, 2
Cold Weather Starting Aid System Use	40°F(4°C) or below
Valve Tappet Clearance Adjustment	

Table 1. Leading Particulars. - Continued

Hot or Cold (Intake)	0.040-0.055 in. (1.0-1.4 mm.)
Hot or Cold (Exhaust)	0.045-0.070 in. (1.1-1.8 mm.)
4. Cooling System:	
Type	Pressurized radiator and pump
Capacity:	
MEP-804A/MEP-814A	13.5 qts. (12.8 liters)
MEP-804B/MEP-814B	11.2 qts. (10.6 liters)
Normal Operating Temperature	170-200°F (77-93°C)
Temperature Indicating System Voltage Rating	24 VDC
5. Lubricating System:	
Type	
MEP-804A/MEP-814A	Full flow, circulating pressure
MEP-804B/MEP-814B	Forced lubrication
Oil Pump Type:	
MEP-804A/MEP-814A	Positive displacement gear
MEP-804B/MEP-814B	Trochoid
Normal Operating Pressure	25-60 psi. (172-414 kPa.)
Oil Filter Type	Full flow, spin-on, replaceable element
Capacity	6 qts. (5.7 liters)
Pressure Indicating System Voltage Rating	24 VDC
6. Fuel System:	
Type of Fuel	DF-1, DF-2, DF-A, JP4, JP5, JP8
Fuel Tank Capacity	14 gal. (53 liters)
Fuel Consumption Rate (50/60 Hz):	
MEP-804A	1.50 gal. (5.7 liters) per hour
MEP-804B	1.20 gal. (4.5 liters) per hour
Fuel Consumption Rate (400 Hz):	
MEP-814A	1.75 gal. (6.6 liters) per hour
MEP-814B	1.40 gal. (5.3 liters) per hour
Auxiliary Fuel Pump:	
Voltage Rating	24 VDC
Delivery Pressure	5.0-6.5 psi. (34.5-65.5 kPa.) (max)
Fuel Level Switch:	
Type	Float
Current	3.0 amps at 6-32 VDC
7. Engine Starting System:	
Batteries	Two 12 volt, connected in series

Table 1. Leading Particulars. - Continued

Starter (MEP-804A/MEP-814A):	
Manufacturer	Hitachi
Model	S25-121
Voltage Rating	24 VDC
Drive Type	Gear reduction
Starter (MEP-804B/MEP-814B):	
Manufacturer	Yanmar
Model	129612-77011
Voltage Rating	24 VDC
Drive Type	Direct drive
Battery Charging Alternator (MEP-804A/MEP-814A):	
Manufacturer	Hitachi
Model	LR220-24
Amperage Rating	20 amps at 24 VDC
Protective Fuse	30 amps
Battery Charging Alternator (MEP-804B/MEP-814B):	
Manufacturer	Yanmar
Model	129900-77240
Amperage Rating	35 amps at 24 VDC
Protective Fuse	30 amps
8. AC Generator:	
Manufacturer	Marathon Electric
Type	Rotating field synchronous
Load Capacity	15 kW
Current Ratings:	50Hz 60Hz 400Hz
120/208 volt connection	43 amps 52 amps 52 amps
240/416 volt connection	21 amps 26 amps 26 amps
Power Factor	0.8
Cooling	Fan cooled
Drive Type	Direct coupling
Duty Classification	Continuous
9. Governing System:	
Load Measuring Unit	
Manufacturer	Technology Research
Model	19310
Governor Control Unit (MEP-804A)	
Manufacturer	Woodward Governor Co.
Model	8270-1002

Table 1. Leading Particulars. - Continued

Governor Control Unit (MEP-804B)	
Manufacturer	Woodward Governor Co.
Model	8270-1096
Governor Control Unit (MEP-814A/MEP-814B)	
Manufacturer	Woodward Governor Co.
Model	8270-1003
10. Protection Devices:	
Low Oil Pressure Switch	
Trip Pressure	15±3 psi. (103.4±20.7 kPa.)
Operating Voltage	12/24 VDC
Current Rating	5 amps
Coolant High Temperature Switch:	
Trip Temperature	220±3.5°F(104±2°C)
Voltage Rating	20 - 32 VDC
Current Rating	7 amps resistive; 4 amps inductive
Overspeed Switch:	
Element Trip and Reset	2200±40 RPM
Voltage Rating	28 VDC
Current Rating	1amp
Overvoltage:	
Trip Point Conditions	155±1 VAC for no less than 200 milliseconds (120 VACcoil winding)
Trip Point	No more than 1.25 seconds after trip conditions exist

END OF WORK PACKAGE

OPERATOR MAINTENANCE

15 kW 50/60 AND 400 Hz SKID MOUNTED, TACTICAL QUIET GENERATOR SET

THEORY OF OPERATION

INTRODUCTION

This work package contains functional descriptions of the generator set and explains how the controls and indicators interact with the system.

ENGINE STARTING SYSTEM

The Engine Starting System (Figure 1), consists of two 12-volt batteries connected in series, a starter, a 24 volt battery charging alternator, a magnetic pickup (for sensing engine speed) and the related switches and relays required for control of the starting system. For engine cranking, battery power is supplied to the starter motor through the starter solenoid which in turn is controlled by the cranking relay. The starter then engages the engine flywheel causing the engine to turn over. For engine starting, the DEAD CRANK switch must be in the NORMAL position, the DC Control power circuit breaker must be pushed in, the Emergency Stop Switch must be in the Out position and the MASTER SWITCH is moved to the START position. The cranking relay is then controlled by a circuit consisting of the start relay and crank disconnect switch. As the engine accelerates to the preset speed
(sensed by the magnetic pickup), the crank disconnect switch opens and deenergizes the cranking relay to stop and disengage the starter. The starting sequence may also be stopped by moving the MASTER SWITCH to OFF. The engine may be cranked without starting by use of the DEAD CRANK switch. With the DEAD CRANK switch in the CRANK position, the cranking relay, starter solenoid and starter motor are energized without activating any other starting or control function. The batteries are charged by the battery charging alternator that is belt driven by the engine. Generator set control system power is also supplied by the battery charging alternator. The BATTERY CHARGE ammeter indicates the charge/discharge rate of the batteries, from -10 AMPS to +20 AMPS, in 5 AMPS increments. Normal operating indication depends on the state of charge in the batteries. A low charge, such as exists immediately after engine starting, will cause a high reading (needle moves toward CHARGE area). When the charge in the batteries has been restored, the indicator moves near zero.

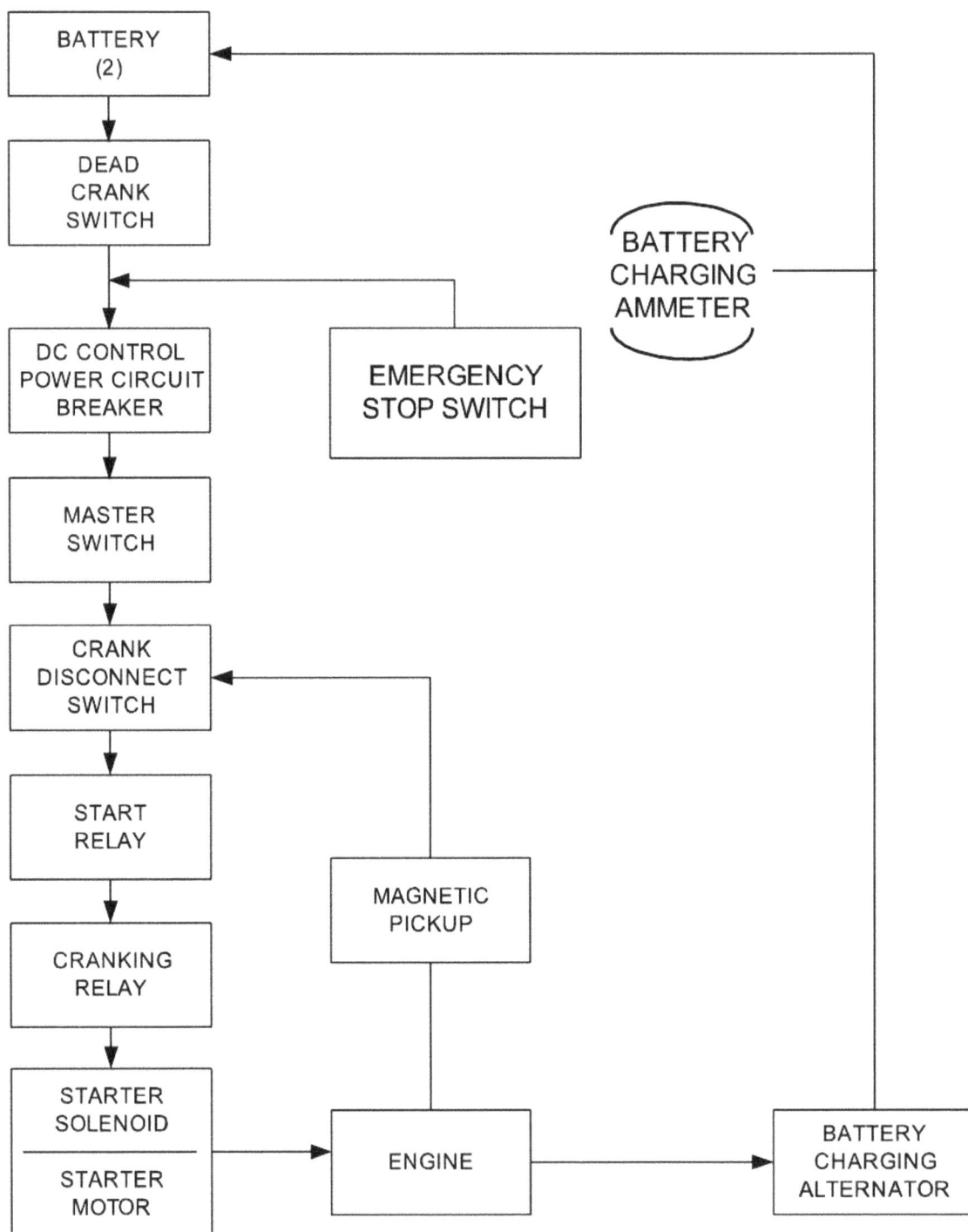

Figure 1. Engine Starting System.

FUEL SYSTEM

The Fuel System (Figure 2), consists of piping, fuel tank, pump fuel filter (MEP-804B/MEP-814B), transfer pump, fuel filter/water separator, engine fuel filter (MEP-804B/MEP-814B), injection pump and injectors. Fuel is drawn from the fuel tank by the transfer pump. After reaching the transfer pump, fuel passes through a fuel filter/water separator where water and small impurities are removed. The fuel then goes to an injection pump where it is pressurized and pushed into the injectors. Through the injectors fuel enters the diesel engine combustion chamber, where it is mixed with air and ignited. The fuel that is not used is returned to the fuel tank via an excess fuel return line.

The Auxiliary Fuel System consists of an external fuel supply, fuel filter, piping, a 24 VDC auxiliary fuel pump and a fuel level float switch. When the MASTER SWITCH is set on PRIME & RUN AUX FUEL it actuates the auxiliary fuel pump and transfers fuel from the external fuel supply to the generator set fuel tank. The fuel level float switch shuts off the auxiliary fuel pump when the generator set fuel tank is full and reactivates the pump as the level drops. The FUEL LEVEL indicator indicates fuel level of generator set fuel tank from (E) empty to (F) full in quarter tank increments.

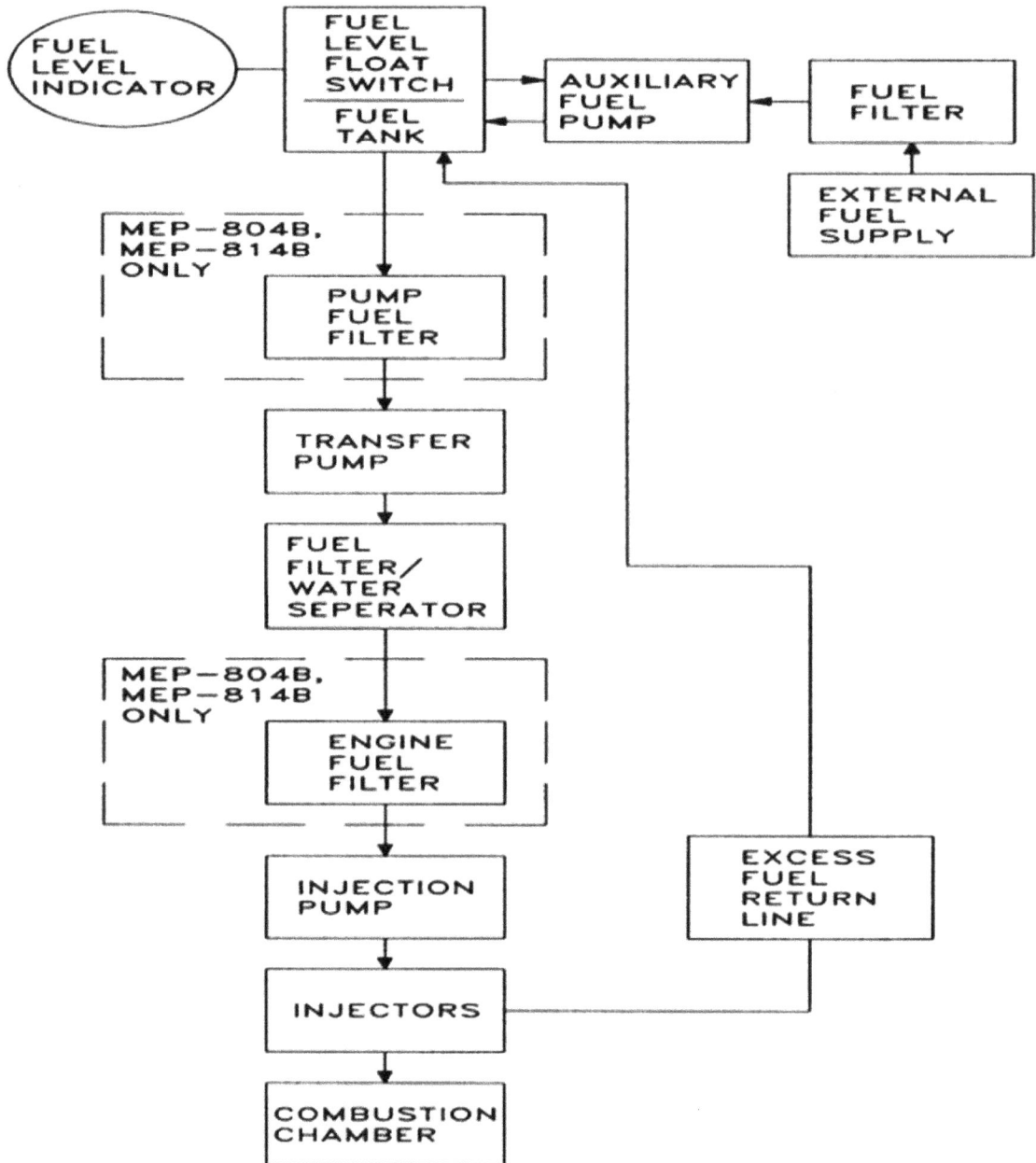

Figure 2. Fuel System.

ENGINE COOLING SYSTEM

The Engine Cooling System (Figure 3) consists of a radiator, hoses, thermostat, water pump, a belt driven fan, and cooling jackets (part of engine). The water pump forces coolant through passages (cooling jackets) in the engine block and cylinder head where the coolant absorbs heat from the engine. When the engine reaches normal operating temperature, the thermostat opens and the heated coolant flows through the upper radiator hose assembly into the radiator. The cooling fan circulates air through the radiator where the coolant temperature is reduced.

A coolant high temperature switch provides automatic shut down in the event that coolant temperature exceeds 220±3.5°F (104±2°C). The COOLANT TEMP indicator indicates the engine coolant temperature, from 120°F to 240°F(48°C to 115°C).

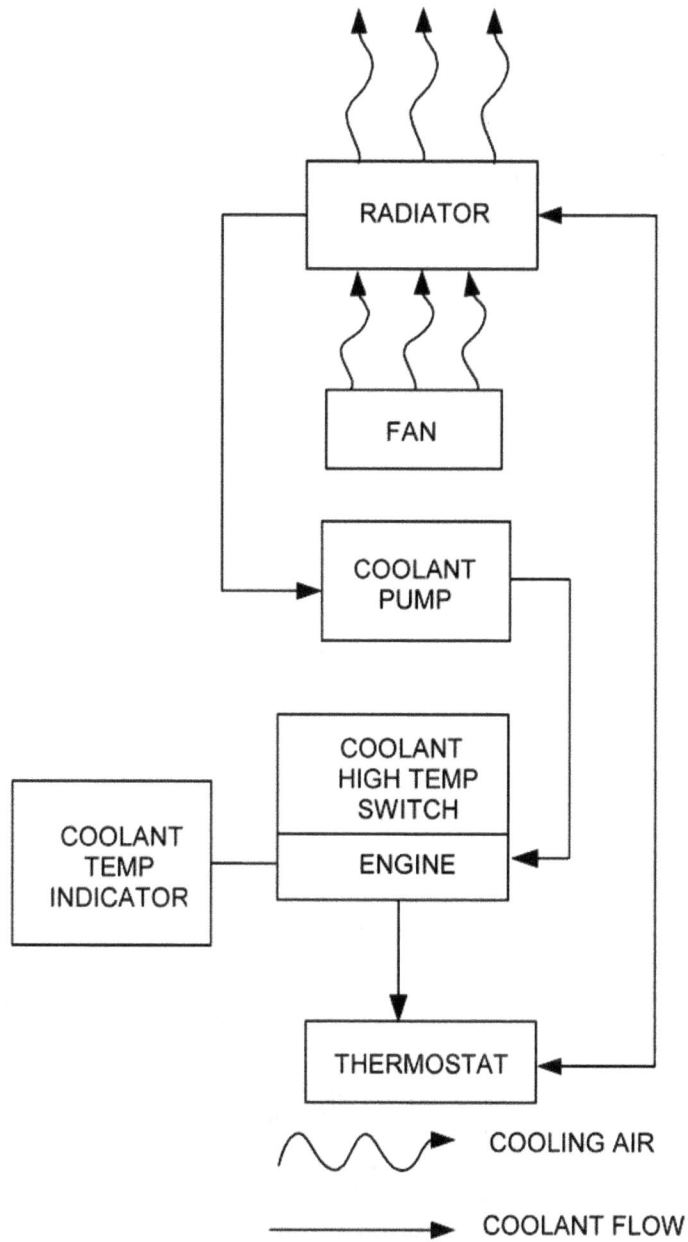

Figure 3. Engine Cooling System.

LUBRICATION SYSTEM

The Lubrication System (Figure 4) consists of an oil pan, dipstick, pump, oil pressure sender, AOAP sample valve, and filter. The oil pan is a reservoir for engine lubricating oil. The dipstick indicates oil level in the oil pan. A pump draws oil from the oil pan and through a screen removing large impurities. The oil then passes through a spin-on type filter where small impurities are removed. From the filter, oil enters the engine and is distributed to the engine's internal moving parts. After passing through the engine, the oil returns to the oil pan. The OIL PRESSURE indicator indicates oil pressure sensed by the oil pressure sender in the engine. The engine will shut off automatically if the oil pressure drops to a dangerously low level. The oil level can be checked with engine running.

Figure 4. Engine Lubrication System.

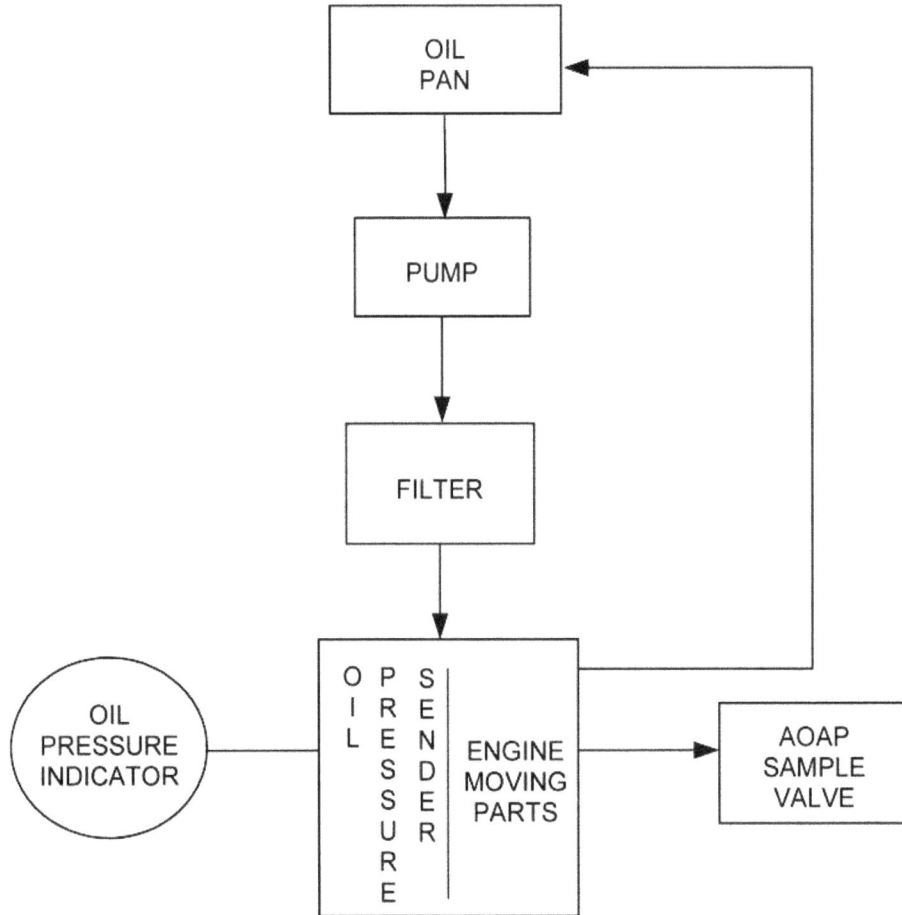

AIR INTAKE AND EXHAUST SYSTEM

The Air Intake and Exhaust System (Figure 5), consists of an air cleaner assembly, intake manifold, glow plugs, exhaust manifold and muffler. Ambient air is drawn into the air cleaner assembly where it passes through the air cleaner element. Airborne dirt is removed and trapped in the element. A restriction indicator, located on the air cleaner assembly housing, displays red when the air cleaner element should be serviced.

On the Isuzu engine (MEP-804A/MEP-814A), filtered air is drawn out of the air cleaner assembly through air intake tubes to the intake manifold where it passes into the engine and is mixed with fuel from the injectors.

On the Yanmar engine (MEP-804B/MEP-814B), filtered air is drawn out of the air cleaner assembly and pressurized by the turbocharger. The pressurized air is then forced into the intake manifold where it passes into the engine and is mixed with fuel from the injectors.

The Isuzu engine (MEP-804A/MEP-814A) uses a positive crankcase ventilation (PCV) valve to circulate engine gases back to the intake manifold, reducing emissions. The Yanmar engine (MEP-804B/MEP-814B) uses a closed crankcase ventilation (CCV) filter for the same purpose. On the Yanmar engine, the gases are circulated back to the turbocharger inlet.

The engine exhaust gases are expelled into the exhaust manifold. On the Isuzu engine (MEP-804A/MEP- 814A), the exhaust manifold channels the gases into the muffler that deadens the sound of the exhaust gases. On the Yanmar engine (MEP-804B/MEP-814B), exhaust gases from the exhaust manifold drive an impeller in the turbocharger. The impeller is mechanically connected to a fan on the intake which pressurizes the intake air prior to entering the engine. The exhaust from the turbocharger is routed to the muffler to deaden the sound of the exhaust. The gases pass from the muffler through the muffler outlet and are vented upward from the generator set housing.

Cold outside temperatures make starting the engine difficult. To improve engine starting, a cold weather starting aid has been provided that is activated when the MASTER SWITCH is in PREHEAT position. The Isuzu engine (MEP-804A/MEP-814A) uses a glow plug in each cylinder to assist in starting. The Yanmar engine (MEP-804B/MEP-814B) uses two intake air pre-heaters to assist in starting.

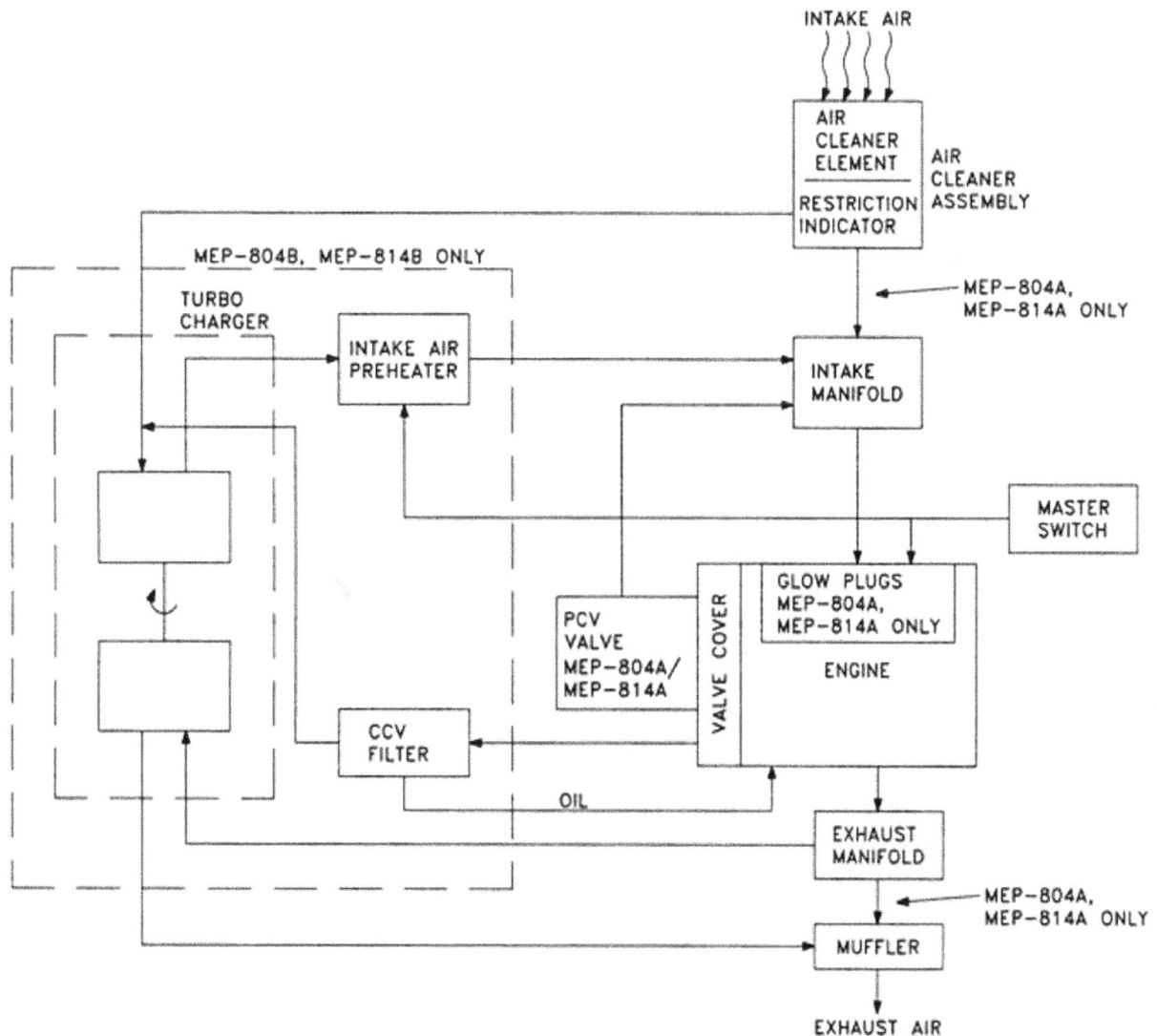

Figure 5. Air Intake and Exhaust System.

OUTPUT SUPPLY SYSTEM

The Output Supply System (Figure 6) consists primarily of the AC generator, the output load terminal board, the voltage reconnection terminal board, the VM-AM transfer switch and the AC circuit interrupter relay. Power created by the AC generator is supplied through the voltage reconnection terminal board and the AC circuit interrupter relay to the output load terminals on the output load terminal board. The voltage reconnection terminal board allows configuration of the generator set for 120/208 volt connections or 240/416 volt connections.

The AC CIRCUIT INTERRUPTER switch closes and opens the AC circuit interrupter relay. This enables or interrupts the power flow between the voltage reconnection terminal board and the output load terminals. The AC circuit interrupter relay is also opened automatically during any of the specified set faults. The voltage regulator senses AC generator output voltage and provides control voltage to the AC generator exciter to maintain the desired AC generator output voltage. The position of the VM-AM transfer switch selects the output load terminals from which current and voltage are measured and are indicated on the AC voltmeter (VOLTS AC) and the ammeter (PERCENT RATED CURRENT).

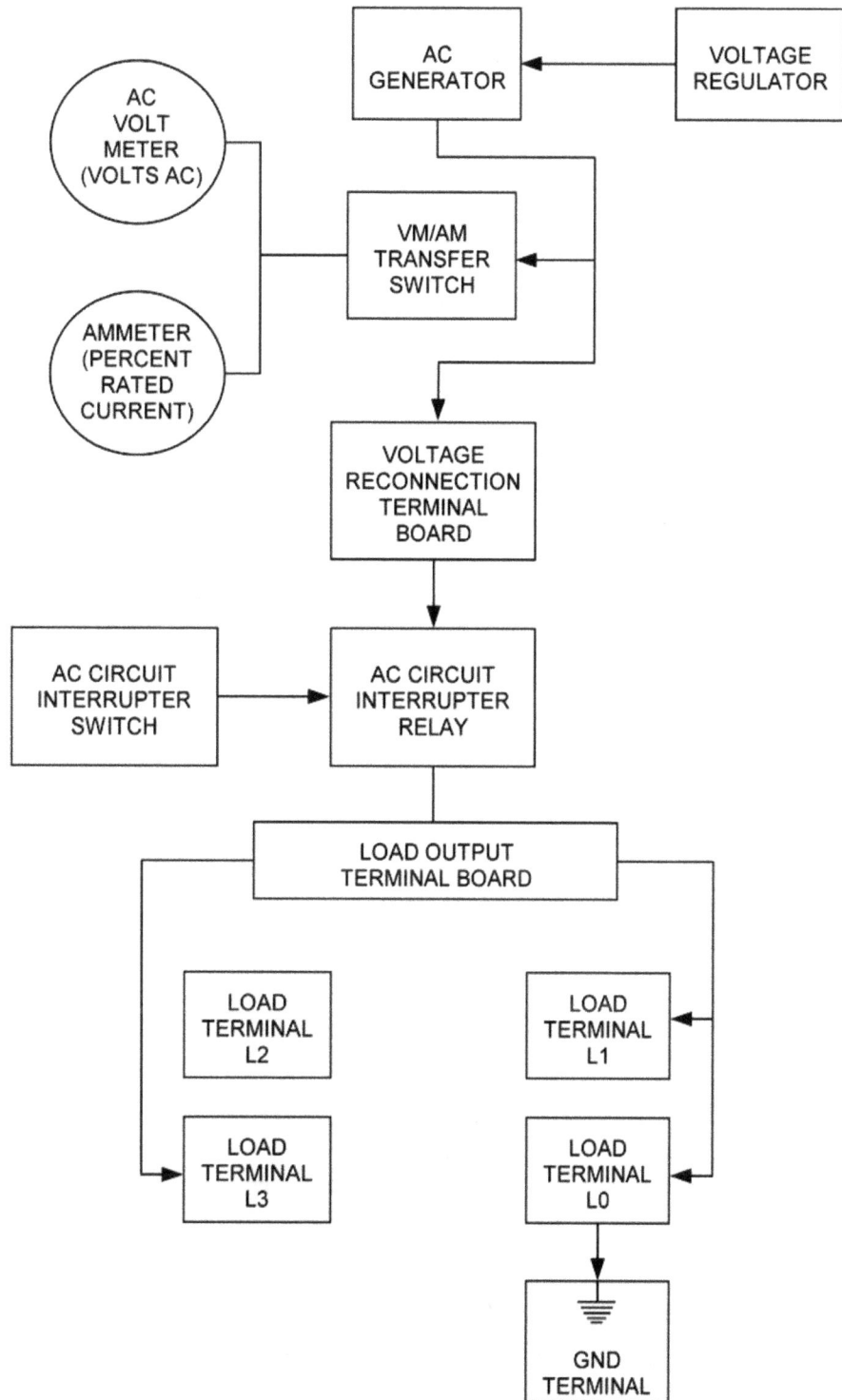

Figure 6. Output Supply System.

WINTERIZATION KIT

See Chapter 6 for Operating Procedures.

END OF WORK PACKAGE

CHAPTER 2

OPERATOR INSTRUCTIONS

FOR

15 kW 50/60 AND 400 Hz SKID MOUNTED, TACTICAL QUIET GENERATOR SET

CHAPTER 2

OPERATOR INSTRUCTIONS

WORK PACKAGE INDEX

OPERATOR MAINTENANCE

15 kW 50/60 AND 400 Hz SKID MOUNTED, TACTICAL QUIET GENERATOR SET
DESCRIPTION AND USE OF OPERATOR CONTROLS AND INDICATORS

GENERAL

This work package describes and illustrates the controls and indicators to ensure proper operation of the generator set.

CONTROL PANEL ASSEMBLY

The control panel assembly contains most of the operating controls and indicators for the generator set. Figure 1 shows the control panel assembly layout and Table 1 describes each control and indicator.

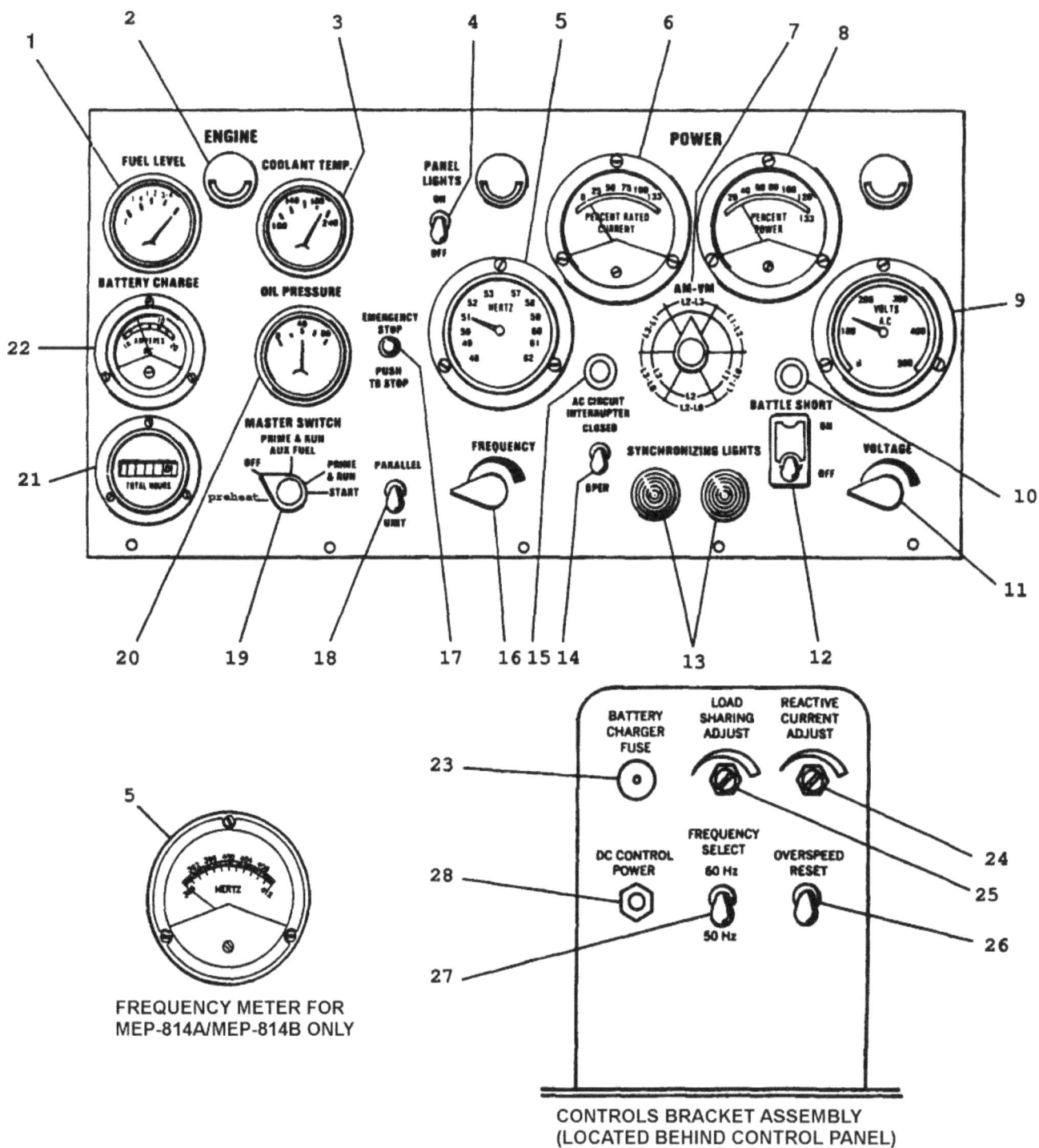

ENGINE

FUEL LEVEL

COOLANT TEMP.

BATTERY CHARGE

OIL PRESSURE

PANEL LIGHTS

POWER

PERCENT RATED CURRENT

PERCENT POWER

AM-VM

VOLTS AC

HERTZ

EMERGENCY STOP
PUSH TO STOP

AC CIRCUIT INTERRUPTER CLOSED

MASTER SWITCH
PRIME & RUN AUX FUEL
OFF
PRIME & RUN
preheat START

PARALLEL
UNIT

FREQUENCY

SYNCHRONIZING LIGHTS

BATTLE SHORT
ON
OFF

VOLTAGE

TOTAL HOURS

FREQUENCY METER FOR
MEP-814A/MEP-814B ONLY

BATTERY CHARGER FUSE

LOAD SHARING ADJUST

REACTIVE CURRENT ADJUST

DC CONTROL POWER

FREQUENCY SELECT
60 Hz
50 Hz

OVERSPEED RESET

CONTROLS BRACKET ASSEMBLY
(LOCATED BEHIND CONTROL PANEL)

Figure 1. Control Panel/Controls Bracket Assembly.

Table 1. Controls and Indicators.

Key	Control/Indicator	Function
1.	FUEL LEVEL indicator	Indicates fuel level.
2.	Panel lights	Illuminates control panel.
3.	COOLANT TEMP indicator	Indicates engine coolant temperature.
4.	PANEL LIGHTS switch	Activates or deactivates panel lights.
5.	FREQUENCY meter (HERTZ)	Indicates generator set output frequency.
6.	Ammeter (PERCENT RATED CURRENT)	Indicates generator set load current as a percent of rated current.
7.	VM-AM transfer switch	Allows selection of current and voltage readings between output load terminals as follows:

Switch Position Voltage Current

L1-LO 120* 240**L1

L2-LO 120* 240**L2

L3-LO 120* 240**L3

L1-L2 208* 416** NONE

L2-L3 208* 416** NONE

L1-L3 208* 416** NONE

AC Reconnection Terminal Board Setting

* 120/208

* 240/416

Key	Control/Indicator	Function
8.	Kilowatt meter (PERCENT POWER)	Indicates generator set output power as a percent of rated power.
9.	AC Voltmeter (VOLTS AC)	Indicates output voltage of generator set.
10.	BATTLE SHORT light	Amber light indicates
11.	VOLTAGE adjust potentiometer	Adjusts generator set voltage.
1.	BATTLE SHORT switch	Bypasses protective devices.
2.	SYNCHRONIZING LIGHTS	Indicates synchronization of units to be paralleled.
1.	AC CIRCUIT INTERRUPTER switch	Opens or closes AC circuit interrupter relay.
1.	AC CIRCUIT INTERRUPTER light	Green light indicates AC circuit interrupter is closed.
1.	FREQUENCY adjust potentiometer	Adjusts frequency of generator set.
1.	EMERGENCY STOP pushbutton	Shuts down generator set.
18.	PARALLEL UNIT switch	Energizes or de-energizes paralleling circuits.

Table 1. Controls and Indicators. - Continued

Key	Control/Indicator	Function
1.	MASTER SWITCH	PREHEAT - Energizes glow plugs.
		OFF - De-energizes all circuits, except panel lights.
		PRIME & RUN AUX FUEL - Energizes generator set run circuits with auxiliary fuel pump operating.
		PRIME & RUN - Energizes generator set run circuits with auxiliary fuel system de-energized.
		START - Energizes starter
20.	OIL PRESSURE indicator	Indicates oil pressure.
21.	Time meter (TOTAL HOURS)	Indicates total engine operating hours.
22.	BATTERY CHARGE ammeter	Indicates charge/discharge rate of batteries.
20. FUSE	BATTERY CHARGER (Located on controls bracket assembly)	on controls bracket assembly) Protects battery charging alternator from overload.
20.	REACTIVE CURRENT ADJUST rheostat (Located on controls bracket assembly)	Adjusts current for load sharing requirements (maintenance personnel only).
25.	LOAD SHARING ADJUST rheostat (Located on controls bracket assembly)	Adjusts power for load sharing requirements (maintenance personnel only).
26. switch	OVERSPEED RESET (Located on controls bracket assembly).	Resets generator set after an overspeed condition (maintenance personnel only).
27. switch	FREQUENCY SELECT (MEP-804A/MEP-804B only) (Located on controls bracket assembly)	Allows selection of 50 Hz or 60 Hz.
28.	DC CONTROL POWER circuit breaker (Located	

MALFUNCTION INDICATOR PANEL

The malfunction indicator panel (Figure 2) is located to the left of the control panel. It contains a series of lights

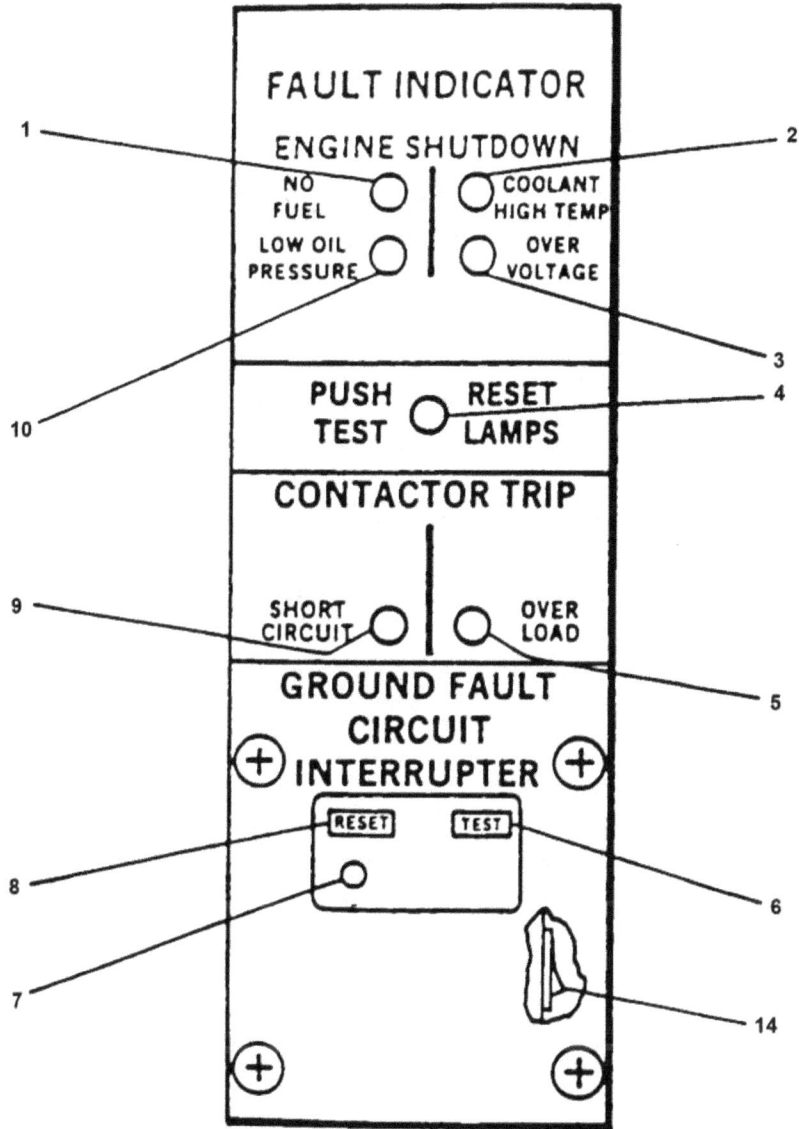

Figure 2. Malfunction Indicator Panel.

Table 2. Controls and Indicators.

Key	Control/Indicator	Function
1.	NO FUEL indicator	Lights when fuel level in fuel tank is below preset level.
2.	COOLANT HIGH TEMP indicator	Lights when engine coolant temperature exceeds 220±3.5°F (104±2°C).
3.	OVERVOLTAGE indicator	Lights when voltage in 120 volt generator coil exceeds 155±1 volts.
4.	OVERSPEED indicator	Lights when engine speed exceeds 2200±40 RPM.
5.	REVERSE POWER indicator	Lights when power flow into generator set exceeds 20±3%of rated current.
6.	OVER LOAD indicator	Lights when current in any phase exceeds 110 % of rated current.

Table 2. Controls and Indicators. - Continued

Key	Control/Indicator	Function
7.	GROUND FAULT CIRCUIT INTERRUPTER TEST pushbutton	Tests GROUND FAULT CIRCUIT INTERRUPTER.
8.	GROUND FAULT CIRCUIT INTERRUPTER indicator	Indicates a ground fault condition.
9.	GROUND FAULT CIRCUIT INTERRUPTER RESET pushbutton	Resets GROUND FAULT CIRCUIT INTERRUPTER.
1.	SHORT CIRCUIT indicator	Lights when generator set output in any phase exceeds 425±25 % of rated current.
1.	UNDERVOLTAGE indicator	Lights when voltage in 120 volt generator coil winding drops below 99±4 VAC.
1.	PUSH TEST RESET LAMPS switch	Tests and resets fault indicator lamps.
1.	LOW OIL PRESSURE indicator	Lights when engine lubrication systems pressure is less than 15±3 psi (103.4±20.7 kPa) during engine operation.
1.	Convenience Receptacle Overload Circuit Breaker (10-amp in-line fuse on generator sets, contract number DAAK01-88-D-0082).	Circuit breaker trips when load on convenience receptacle exceeds 10 amps (fuse blows on generator sets, contract number DAAK01-88-D- 0082).

END OF WORK PACKAGE

OPERATOR MAINTENANCE

15 kW 50/60 AND 400 Hz SKID MOUNTED, TACTICAL QUIET GENERATOR SET
OPERATION UNDER USUAL CONDITIONS

INITIAL SETUP:

Tools and Special Tools
Generator Mechanical Tool Kit

Personnel Required
One

Equipment Condition
Generator set grounded, off & operational

Materials/Parts
Ground rod assembly, ground conductor/cable

References
FM 5-424
WP 0004
WP 0005, Starting/Stopping Procedure
WP 0011

GENERAL

This work package provides information and guidance for generator set operation under normal conditions, refer to FM 2031.

ASSEMBLY AND PREPARATION FOR USE

Installation of Ground Rod

WARNING

Metal jewelry can conduct electricity and become entangled in generator set components. Remove all metal jewelry when working on generator set. Failure to comply can cause injury or death to personnel.

WARNING

Do not wear loose clothing when performing checks, services and maintenance. Loose clothing may be entangled in generator set components. Failure to comply can cause injury or death to personnel.

WARNING

High voltage is produced when the generator set is in operation. Never attempt to start the generator set unless it is properly grounded. Failure to comply can cause injury or death to personnel.

1. Insert ground cable (Figure 1, Item 2) through slot on load output terminal board terminal marked GND (1). Tighten terminal nut.

2. Connect coupling (5) to ground rod (4) and screw driving stud (3) into coupling (5). Make sure that driving stud (3) seats on ground rod (4).

3. Drive ground rod into ground until coupling is just above surface.

4. Remove driving stud and install another section of ground rod.

5. Install another coupling (5) and driving stud (3). Drive ground rod down until new coupling is just above ground surface.

6. Repeat steps 4 and 5 until ground rod has been driven eight feet or deeper, providing an effective ground.

7. Connect clamp (6) and ground cable (2) to ground rod (4) and tighten clamp screw.

Installation of Load Cables

WARNING

Metal jewelry can conduct electricity and become entangled in generator set components. Remove all metal jewelry when working on generator set. Failure to comply can cause injury or death to personnel.

WARNING

Do not wear loose clothing when performing checks, services and maintenance. Loose clothing may be entangled in generator set components. Failure to comply can cause injury or death to personnel.

WARNING

High voltage is produced when the generator set is in operation. Never attempt to connect or disconnect load cables while the generator set is running. Failure to comply can cause injury or death to personnel.

WARNING

Dangerous voltage exists on live circuits. Always observe precautions and never work alone. Failure to comply with this warning can cause injury or death to personnel.

WARNING

High voltage is produced when this generator set is in operation. Make sure unit is completely shut down and free of any power source before attempting any repair or maintenance on the unit. Failure to comply can cause injury or death to personnel.

CAUTION

Do not connect the load cables to the convenience receptacle. Failure to observe this caution can result in damage to the generator set.

1. Shutdown generator set.
2. Select required output terminals from Table 1.
3. Open output load terminal door.

WARNING

Jumper will not be removed unless equipment being powered specifically required an isolated ground (floating ground). Failure to comply with this warning can cause injury or death to personnel.

4. Ensure that jumper is securely fastened between LO and ground.
5. Using terminal nut wrench (Figure 2, Item 3) loosen terminal nuts (1) on terminals (2) selected in Step 2.
6. Using terminal nut wrench (Figure 2, Item 3) loosen terminal nuts (1) on terminals (2) selected in Step 2.
7. Insert ends of load cables through load cable entrance box. Insert ends of cables into slots of load terminal studs (2).
8. Tighten load terminal nuts (1).
9. Secure wrench (3) in bracket inside load terminal board door, and close door.

CAUTION

When using single phase connections, always attempt to balance loads between terminals (do not connect all loads between one terminal and LO). Failure to observe this caution can result in damage to generator set.

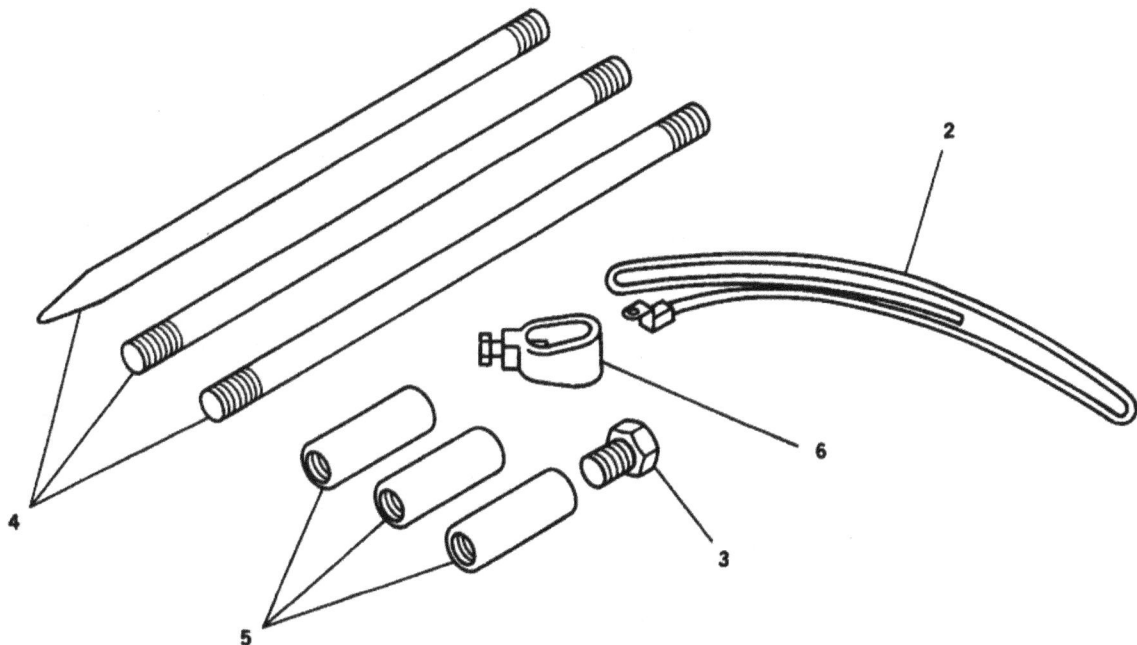

Figure 1. Grounding Connections.

Table 1. Load Terminal, AC Reconnection Board and VM-AM Transfer Switch Selection.

RECONNECTION BOARD POSITION	TERMINALS	VM-AM TRANSFER SWITCH POSITION	VOLTAGE READING	CURRENT READING (TERMINAL)
120/208	L1, L2, L3, LO 3 PHASE. (SINGLE PHASE LOADS CAN BE SERVED USING ANY TERMINAL TO LO)	L1 - LO	120 VOLTS	L1
		L2 - LO	120 VOLTS	L2
		L3 - LO	120 VOLTS	L3
		L1 - L2	208 VOLTS	NONE
		L2 - L3	208 VOLTS	NONE
		L3 - L1	208 VOLTS	NONE
240/416	L1, L2, L3, LO 3 PHASE. (SINGLE PHASE LOADS CAN BE SERVED USING ANY TERMINAL TO LO)	L1 - LO	240 VOLTS	L1
		L2 - LO	240 VOLTS	L2
		L3 - LO	240 VOLTS	L3
		L1 - L2	416 VOLTS	NONE
		L2 - L3	416 VOLTS	NONE
		L3 - L1	416 VOLTS	NONE

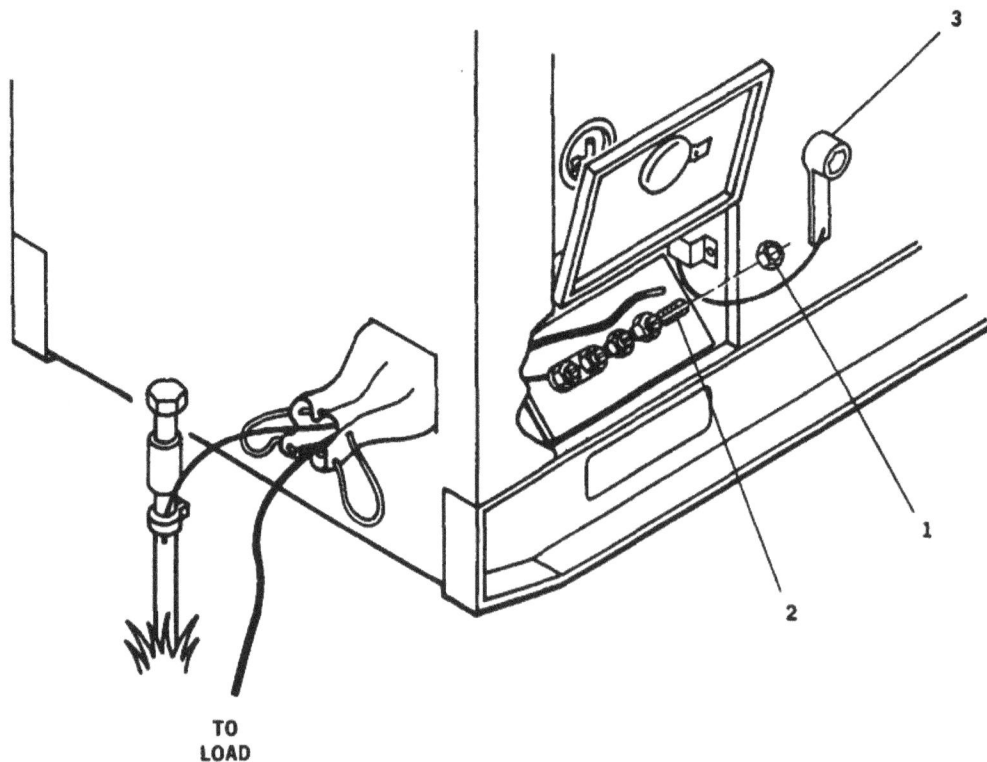

TO
LOAD

Figure 2. Installation of Load Cables.

END OF TASK

DAILY CHECKS, INITIAL ADJUSTMENTS, AND SELF-TEST

Daily Checks

Perform all Before PMCS, refer to WP 0011, Table 1.

Initial Adjustments

1. Place DEAD CRANK switch in NORMAL position.
2. Push DC CONTROL POWER circuit breaker in.
3. Place FREQUENCY SELECT switch to required position (50/60 Hz).
4. Ensure voltage reconnection terminal board is positioned to match voltage requirements. If voltage reconnection terminal board must be changed, notify next higher maintenance level.
5. Place VM-AM transfer switch in a position corresponding to output terminal load connections, refer to Table 1.
6. Place PARALLEL UNIT switch in UNIT position.
7. Pull out Emergency Stop Switch.

END OF TASK

Self Test

1. Place MASTER SWITCH to PRIME& RUN position.
2. Push PRESS TO TEST pushbutton on malfunction indicator panel. Ensure all indicator lights are lit. When PRESS TO TEST pushbutton is released, all lights should go out.
3. Press BATTLE SHORT press to test light on the control panel assembly. Ensure indicator light is lit. When press to test light is released, light should go out.
4. Press AC CIRCUIT INTERRUPTER press to test light on the control panel assembly. Ensure indicator light is lit. When press to test light is released light should go out.

END OF TASK

OPERATING PROCEDURES

WARNING

Metal jewelry can conduct electricity and become entangled in generator set components. Remove all metal jewelry when working on generator set. Failure to comply can cause injury or death to personnel.

WARNING

Do not wear loose clothing when performing checks, services and maintenance. Loose clothing may be entangled in generator set components. Failure to comply can cause injury or death to personnel.

WARNING

High voltage is produced when the generator set is in operation. Never attempt to start the generator set unless it is properly grounded. Failure to comply can cause injury or death to personnel.

WARNING

High voltage is produced when the generator set is in operation. Never attempt to connect or disconnect load cables while the generator set is running. Failure to comply can cause injury or death to personnel.

WARNING

Exhaust discharge contains deadly gases including carbon monoxide. Do not operate generator set in an enclosed area unless exhaust discharge is properly vented outside. Failure to comply can cause injury or death to personnel.

NOTE

If generator set is to be operated in parallel with another unit, refer to Parallel Unit Operation.

Starting Procedure

WARNING

Metal jewelry can conduct electricity and become entangled in generator set components. Remove all metal jewelry when working on generator set. Failure to comply can cause injury or death to personnel.

WARNING

Do not wear loose clothing when performing checks, services and maintenance. Loose clothing may be entangled in generator set components. Failure to comply can cause injury or death to personnel.

WARNING

High voltage is produced when the generator set is in operation. Never attempt to start the generator set unless it is properly grounded. Failure to comply can cause injury or death to personnel.

WARNING

Operating the generator set exposes personnel to a high noise level. Hearing protection must be worn when operating or working near the generator set when the generator set is running. Failure to comply can cause hearing damage to personnel.

CAUTION

Do not crank engine in excess of fifteen seconds. Allow starter to cool at least fifteen seconds between attempted starts. Failure to observe this caution could result in damage to the starter.

NOTE

At temperatures below 40°F(4°C) it may be necessary to use the Cold Weather Starting Aid.

NOTE

Ensure all generator set access doors, except control panel access door, are closed.

1. At temperatures below 40°F(4°C) turn MASTER SWITCH to PREHEAT for 30 seconds.
2. Rotate MASTER SWITCH to START position.
3. Hold MASTER SWITCH in START position until oil pressure reaches at least 25 psi (172 kPa), voltage has increased to its approximate rated value, and engine has reached stable operating speed.
4. Release MASTER SWITCH to PRIME AND RUN position.

5. If operating with an auxiliary fuel source, rotate MASTER SWITCH to PRIME AND RUN AUX FUEL position.

NOTE

Warm up engine without load for five minutes. (If required, load can be applied immediately).

6. Check COOLANT TEMP [170-200°F (77-93°C)] and OIL PRESSURE [25-60 psi (172-414 kPa)] indicators for normal readings.

7. Turn VOLTAGE and FREQUENCY adjust potentiometers to required values for voltage and frequency.

8. Press GROUND FAULT CIRCUIT INTERRUPTER TEST pushbutton. Ensure indicator window is clear. Press RESET pushbutton and ensure indicator is red.

9. Place AC CIRCUIT INTERRUPTER switch to CLOSED position.

10. Ensure voltage and frequency are still at rated values. Adjust if necessary.

11. Rotate VM-AM transfer switch to each phase position while observing ammeter (PERCENT RATED CURRENT). If more than rated load is indicated in any phase, reduce load.

12. Check kilowattmeter (PERCENT POWER). If indication is more than 100 percent rated load, reduce load.

13. Perform all During OPERATION PMCS requirements in accordance with WP 0011, Table 1.

END OF TASK

Stopping Procedure

1. Place AC CIRCUIT INTERRUPTER switch in OPEN position.

2. Allow generator set to operate five minutes with no load applied.

3. Place MASTER SWITCH in OFF position.

4. Perform all AFTER OPERATION PMCS requirements in accordance with WP 0011, Table 1.

5. Place DEAD CRANK switch in OFF position.

END OF TASK

PARALLEL UNIT OPERATION (LOAD SHARING)

CAUTION

Ensure generator sets are the same size and mode before attempting parallel operation.

General

The following of parallel operation will be used to share the load between two generator sets. Refer to Figure 2-1 for location of operator controls and indicators mentioned below and Figure 2-5 for proper paralleling configuration.

Pre-Operation

WARNING

Metal jewelry can conduct electricity and become entangled in generator set components. Remove all metal jewelry when working on generator set. Failure to comply can cause injury or death to personnel.

WARNING

Do not wear loose clothing when performing checks, services and maintenance. Loose clothing may be entangled in generator set components. Failure to comply can cause injury or death to personnel.

WARNING

High voltage is produced when this generator set is in operation. Make sure unit is completely shut down and free of any power source before attempting any repair or maintenance on the unit. Failure to comply can cause injury or death to personnel.

WARNING

High voltage is produced when the generator set is in operation. Never attempt to connect or disconnect load cables while the generator set is running. Failure to comply can cause injury or death to personnel.

WARNING

Dangerous voltage exists on live circuits. Always observe precautions and never work alone. Failure to comply with this warning can cause injury or death to personnel.

1. Ensure that load requirement is equal or below the combined rated capacity of the two generator sets.

WARNING

High voltage is produced when the generator set is in operation. Never attempt to start the generator set unless it is properly grounded. Failure to comply can cause injury or death to personnel.

20.   Determine voltage requirements of load and position voltage reconnection terminal boards of the two generator sets to the required voltage connection. Ensure FREQUENCY SELECT switch (MEP-804A/MEP-804B) for both generator sets are positioned for the same load requirements.

3. Identify one generator set as No. 1 and the other as No. 2.

4. Remove paralleling cable from storage box located inside battery compartment access door.

5. Connect the paralleling cable between the two generator sets. Connect the generator sets to the load observing the proper phase polarity.

END OF TASK

Operation

CAUTION

Do not close the AC CIRCUIT INTERRUPTER switch on either of the generator sets, nor close the load contactor at load until specifically directed to do so. Closing any of these devices at any other time may severely damage one or both of the generator sets.

1. Start each generator set, refer to Starting Procedure.

2. Rotate both VOLTAGE adjust potentiometers to obtain the same voltage indication on each set.

3. Rotate both FREQUENCY adjust potentiometers to obtain the same frequency indication on both sets. Ensure load contactor at load is open.

4. Position and hold AC CIRCUIT INTERRUPTER switch, on generator set No. 1, to CLOSED until indicator lights.

5. Place the UNIT-PARALLEL switch on both units in PARALLEL position.

WARNING

Power is available to the convenience receptacle when the generator set is running. Avoid accidental contact. Failure to comply may cause injury or death to personnel.

CAUTION

If synchronizing lights on generator set No. 2 do not glow bright and dark in unison, the phasing is wrong. Shut down generator sets and check that load cables are connected properly. Failure to observe this caution can result in damage to generator sets.

6. Observe synchronizing lights on generator set No. 2. The lights should be glowing bright and dark in unison.

7. Adjust frequency of generator set No. 2 until synchronizing lights glow bright and dark in unison at 2 to 3 second intervals.

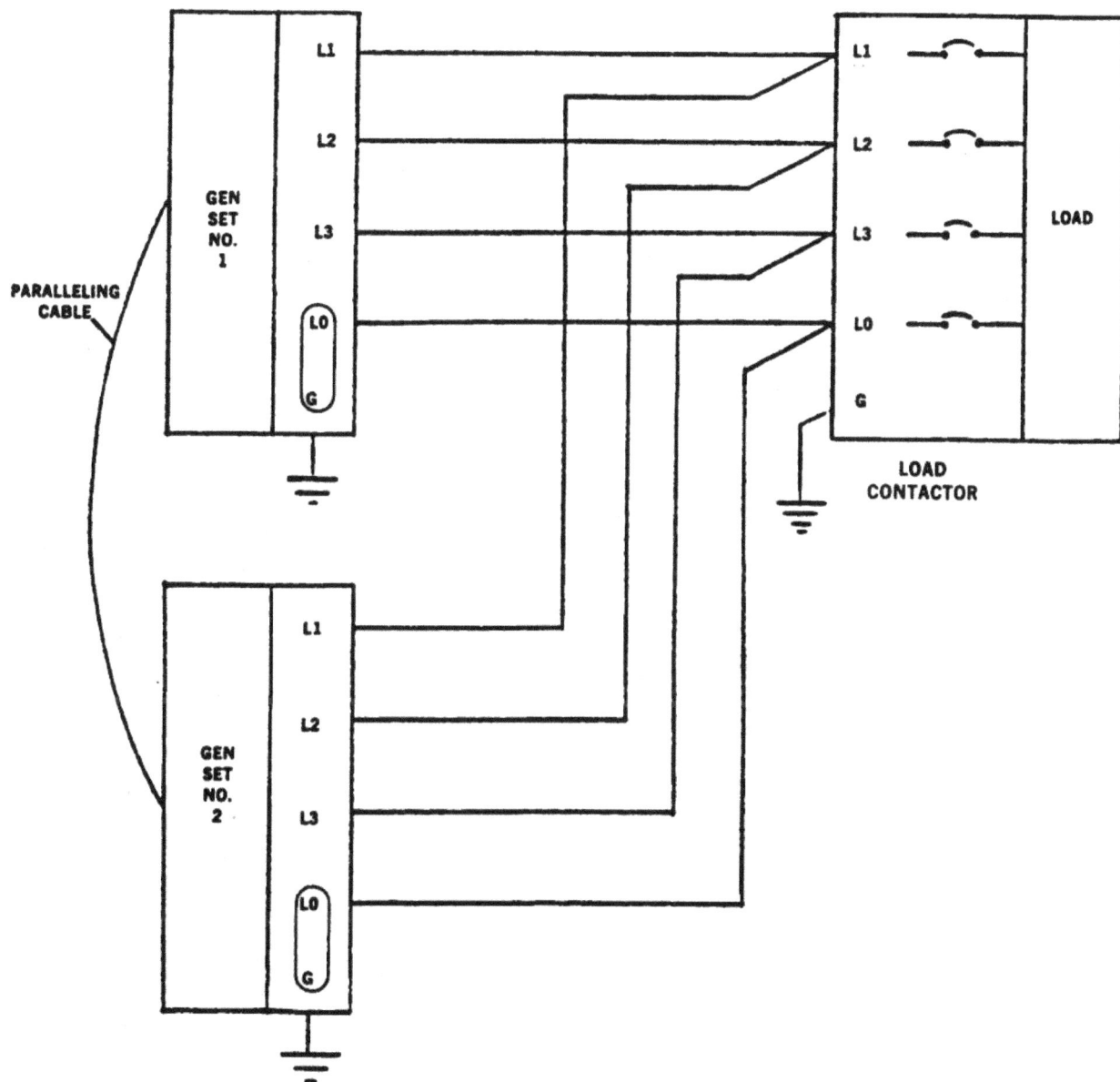

Figure 3. Parallel Operation Setup.

CAUTION

Check that load contactor at load is open before attempting to place generator sets on line. Failure to observe this caution can result in damage to generator sets.

8. When both synchronizing lights are dark, position and hold AC CIRCUIT INTERRUPTER switch of generator set No. 2 to the CLOSED position until indicator lights.

NOTE

The generator sets are now operating in parallel with no load.

9. Rotate FREQUENCY adjust potentiometer of generator set No. 1 until kilowattmeter (PERCENT POWER) indicates approximately "0".

10. Rotate the VOLTAGE adjust potentiometer of generator set No. 1 until ammeter (PERCENT RATED CURRENT) reads approximately "0".

11.   Close the load contactor at the load.

NOTE

If the REVERSE POWER indicator of either generator set lights, and the AC Circuit Interrupter relay opens, open the load contactor at load and resynchronize the generator sets. (Repeat the necessary steps 4 through 11 above.)

12. Compare ammeter (PERCENT RATED CURRENT) readings of both generator sets. If readings are not within 10%, notify next higher level of maintenance.

13.   Compare kilowattmeter (PERCENT POWER) readings of both generator sets. If readings are not within 10%, notify next higher level of maintenance.

END OF TASK

Removal from Parallel Operation

WARNING

Metal jewelry can conduct electricity and become entangled in generator set components. Remove all metal jewelry when working on generator set. Failure to comply can cause injury or death to personnel.

WARNING

Do not wear loose clothing when performing checks, services and maintenance. Loose clothing may be entangled in generator set components. Failure to comply can cause injury or death to personnel.

WARNING

High voltage is produced when the generator set is in operation. Never attempt to connect or disconnect load cables while the generator set is running. Failure to comply can cause injury or death to personnel.

WARNING

If necessary to move a generator set that has been operating in parallel with another generator set, shut down all generator sets prior to removing load cables or ground. Failure to comply can cause injury or death to personnel by electrocution.

CAUTION

Prior to removal of generator set from parallel operation, make sure load does not exceed full load rating of generator set remaining on line. Failure to observe this caution can result in damage to generator set.

1. Position AC CIRCUIT INTERRUPTER switch to OPEN until indicator goes out.

2. Return UNIT-PARALLEL switch to UNIT position.

3. Refer to Stopping Procedure to stop generator set.

END OF TASK

DECALS AND INSTRUCTION PLATES

There are identification and instruction plates on the generator set. Figure 4 through Figure 21 show the location and contents of each plate on the generator set.

GROUNDING STUD PLATE
VOLTAGE CONNECTION CAUTION PLATE
SCHEMATIC DIAGRAM
WIRING DIAGRAM

RIGHT SIDE

BATTERY CONNECTION INSTRUCTION

FRONT

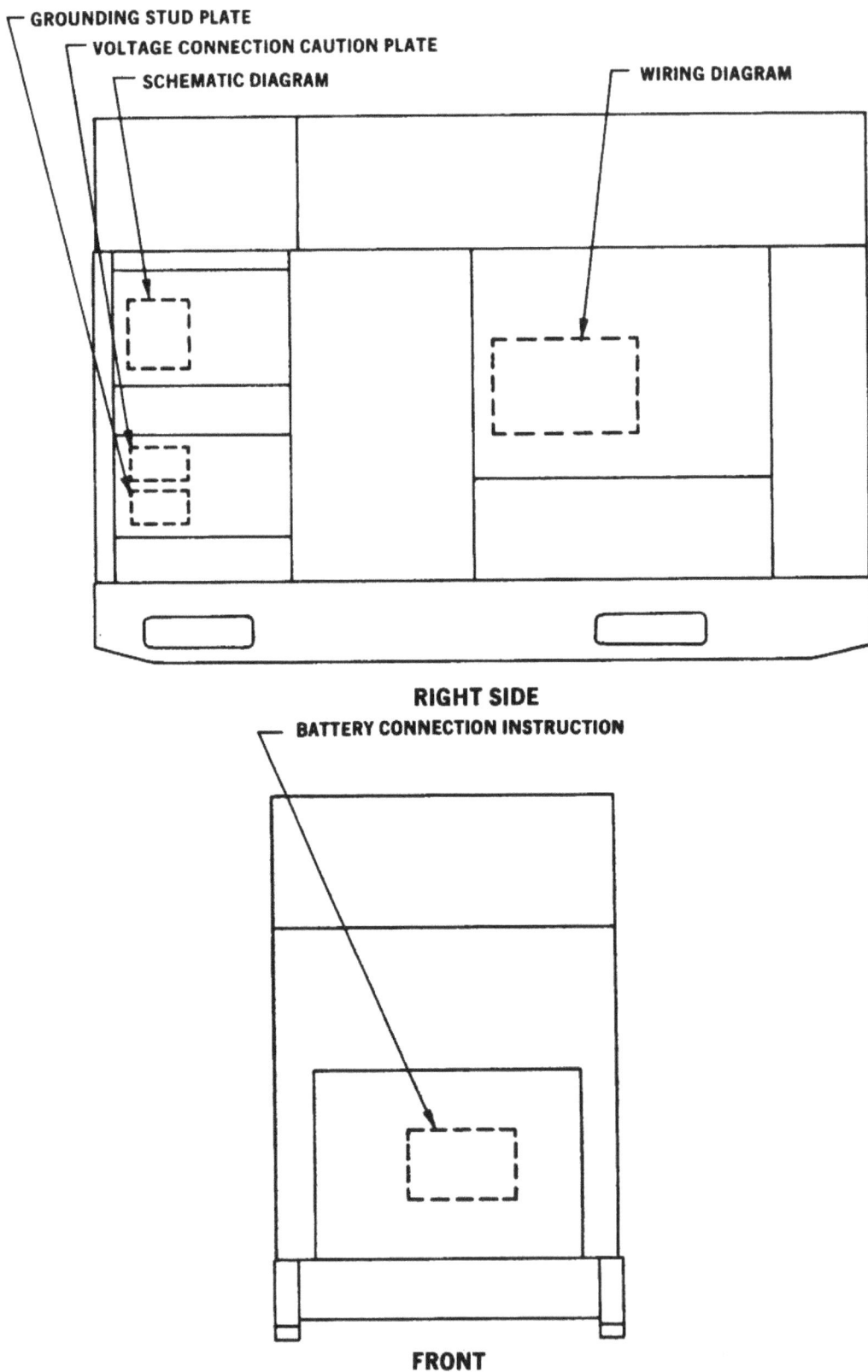

Figure 4. Operating Instructions Plates (Front and Right Side).

IDENTIFICATION PLATE

SET RATING PLATES

FUEL SYSTEM DIAGRAM

LIFTING AND TIEDOWN DIAGRAM PLATE

EXTERNAL FUEL SUPPLY PLATE

LEFT SIDE

SLAVE RECEPTACLE

PARALLELING RECEPTACLE PLATE

CONVENIENCE RECEPTACLE PLATE

DIAGNOSTIC PLATE

OPERATING INSTRUCTION PLATE

REAR

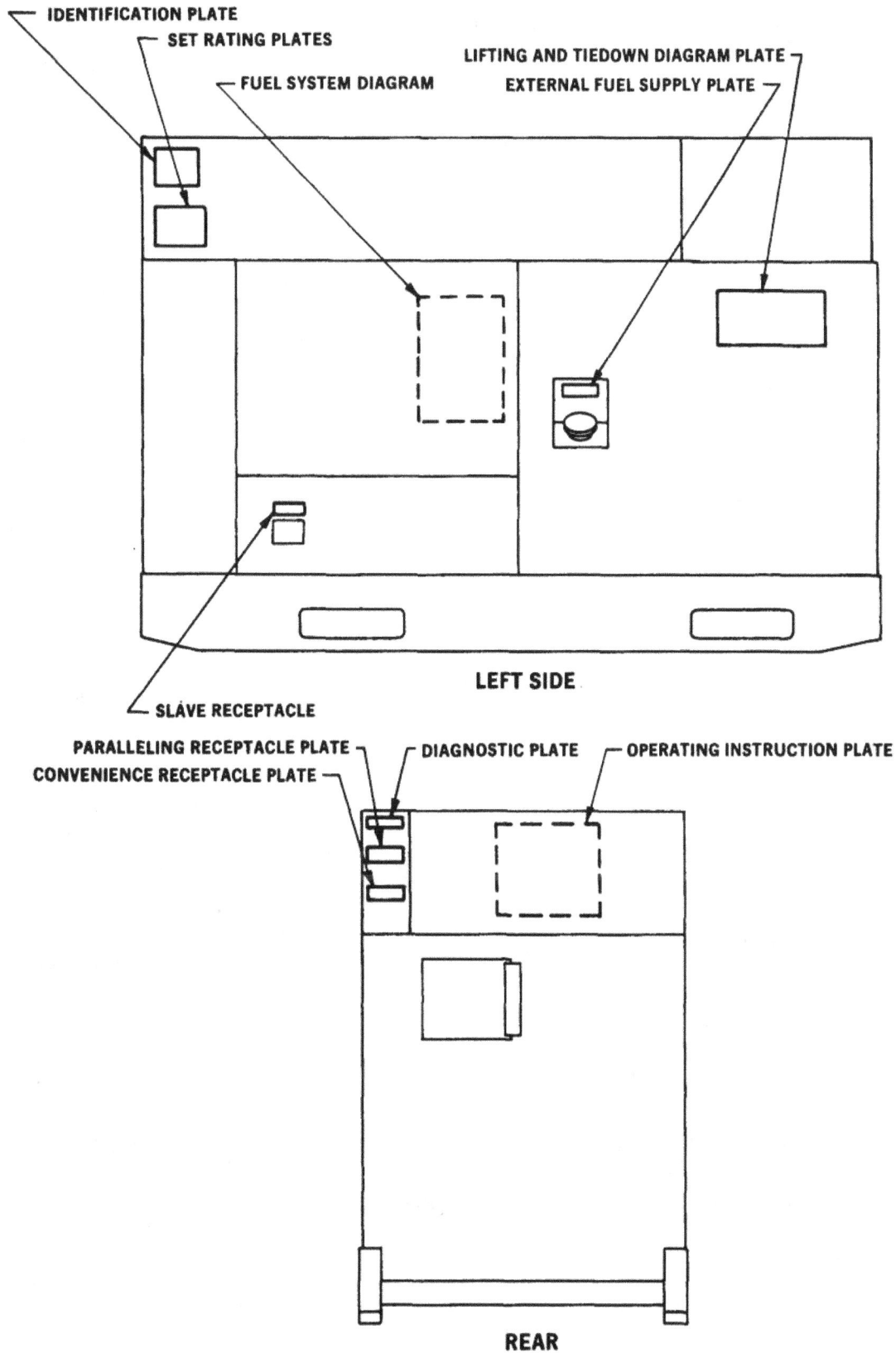

Figure 5. Operating Instructions Plates (Rear and Left Side).

OPERATING

WARNING A. TO AVOID SHOCK HAZARD SET FRAME MUST BE GROUNDED. CONNECT AWG. NO. 8 WIRE OR LARGER FROM GROUND TERMINAL (GND) TO EARTH GROUND

  B. BATTERY NEGATIVE TERMINAL IS CONNECTED TO GROUND.

  C. IDLING OF THE ENGINE AT SPEEDS SLOWER THAN THOSE ATTAINABLE THROUGH THE CONTROLS MAY RESULT IN DAMAGE TO ELECTRICAL COMPONENTS.

1. PRESTART CHECKS

  A. CHECK RADIATOR COOLANT, ENGINE LUBE OIL, FUEL AND BATTERY ELECTROLYTE LEVEL.

  B. CHECK FUEL-WATER SEPARATOR, DRAIN WATER IF PRESENT.

  C. PLACE CONTROL SWITCHES TO OFF OR EQUIVALENT POSITION.

2. NORMAL START (TEMPERATURE ABOVE -25° F)

  A. CRANK THE ENGINE BY PLACING THE MASTER SWITCH IN THE START POSITION. DO NOT CRANK FOR CONTINUOUS PERIODS LONGER THAN 15 SECONDS.

  B. AT TEMPERATURES BELOW APPROXIMATELY 40° F IT MAY BE NECESSARY TO USE THE PRE-HEAT GLOW PLUGS. HOLD THE MASTER SWITCH IN THE PRE-HEAT POSITION FOR 30 SECONDS MAXIMUM PRIOR TO STEP C.

  C. HOLD MASTER SWITCH IN START POSITION UNTIL OIL PRESSURE BUILDS UP TO AT LEAST 25 PSI THEN RELEASE TO PRIME & RUN POSITION.

  D. ADJUST VOLTAGE AND FREQUENCY TO PROPER VALUES. IF NECESSARY, RESET FAULT INDICATOR LIGHTS.

  E. UNDER NORMAL CONDITIONS RUN ENGINE AT NO LOAD FOR 5 MINUTES FOR WARM UP. IF REQUIRED, LOAD CAN BE APPLIED IMMEDIATELY.

  F. CLOSE THE AC CIRCUIT INTERRUPTER BY PLACING THE AC CIRCUIT INTERRUPTER SWITCH IN THE CLOSED POSITION.

3. STOPPING THE SET

  A. REMOVE LOAD BY PLACING THE AC CIRCUIT INTERRUPTER SWITCH IN OPEN POSITION.

  B. ALLOW ENGINE TO OPERATE FOR APPROXIMATELY 5 MINUTES AT NO LOAD.

  C. STOP UNIT BY PLACING MASTER SWITCH IN OFF POSITION.

INSTRUCTIONS

4. PARALLEL OPERATION (2 OR MORE LIKE SETS)

  A. MAKE CONNECTIONS BETWEEN SETS AND LOAD AS DESCRIBED IN THE OPERATING MANUAL.

  B. CONNECT PARALLELING CABLE.

  C. START UNITS NO. 1 AND NO. 2 PER STARTING INSTRUCTIONS.

  D. ADJUST VOLTAGE AND FREQUENCY TO DESIRED VALUE (MUST BE SAME ON BOTH UNITS).

  E. CLOSE AC CIRCUIT INTERRUPTER ON UNIT NO. 1 ONLY.

  F. PLACE UNIT-PARALLEL SWITCH ON BOTH SETS IN PARALLEL POSITION.

  G. OBSERVE SYNCHRONIZING LIGHTS ON UNIT NO. 2 WHICH SHOULD BE ALTERNATELY GLOWING DARK AND BRIGHT IN UNISON. ADJUST FREQUENCY OF UNIT NO. 2 SLIGHTLY AS NECESSARY TO CAUSE LIGHTS TO SLOWLY GLOW BRIGHT AND DARK IN UNISON.

  H. WHEN BOTH LAMPS ARE DARK, CLOSE THE AC CIRCUIT INTERRUPTER ON UNIT NO. 2 (THE UNITS ARE NOW OPERATING IN PARALLEL AND SHOULD APPROXIMATELY DIVIDE KILOWATT LOAD AND CURRENT EQUALLY.)

5. REFER TO APPLICABLE TECHNICAL MANUAL FOR ADDITIONAL INFORMATION ON MAINTENANCE AND TROUBLESHOOTING PROCEDURES.

SERVICE INSTRUCTIONS

FUEL AND OIL				COOLANT	
AMBIENT TEMPERATURE	DIESEL FUEL	LUBRICATING OIL		AMBIENT TEMPERATURE	RADIATOR COOLANT
+20F TO +120F	VV-F-800 GR DF-2	MIL-L-2104C OE HDO-30		+40F TO +120F	WATER MIL-A-53009
0F TO +20F	VV-F-800 GR DF-1	MIL-L-2104C OE HDO-10		-25F TO +120F	WATER MIL-A-46153
-25F TO 0F	VV-F-800 GR DF-1	MIL-L-46167		-25F TO +120F	MIL-A-11755
-25F TO 0F	VV-F-800 GR DF-A	MIL-L-46167			

SYSTEM CAPACITY

FUEL TANK	LUBRICATING OIL			COOLING SYSTEM	
	CRANKCASE		FILTERS	RADIATOR AND OVERFLOW	BLOCK
	FULL	LOW	FILTERS DRAIN TO CRANKCASE		
14 GALLONS	6 QTS.	5 QTS.	-0- QTS.	8 QTS.	5.5 QTS.

NOTE: FOR OPERATION USING JP4, JP5, OR JP8 FUEL REFER TO APPLICABLE OPERATING INSTRUCTION MANUAL

30554-88-22078

Figure 6. Operating Instructions Plate - MEP-804A/MEP-814A.

OPERATING

WARNING:
A. TO AVOID SHOCK HAZARD SET FRAME MUST BE GROUNDED. CONNECT AWG. NO. 6 WIRE OR LARGER FROM GROUND TERMINAL (GND) TO EARTH GROUND.

B. BATTERY NEGATIVE TERMINAL IS CONNECTED TO GROUND.

C. IDLING OF THE ENGINE AT SPEEDS SLOWER THAN THOSE ATTAINABLE THROUGH THE CONTROLS MAY RESULT IN DAMAGE TO ELECTRICAL COMPONENTS.

1. PRESTART CHECKS

A. CHECK RADIATOR COOLANT, ENGINE LUBE OIL, FUEL, AND BATTERY ELECTROLYTE LEVEL.

B. CHECK FUEL-WATER SEPARATOR, DRAIN WATER IF PRESENT.

C. PLACE CONTROL SWITCHES TO OFF OR EQUIVALENT POSITION.

D. PLACE FREQUENCY SELECTOR SWITCH LOCATED WITHIN THE CONTROL BOX IN DESIRED POSITION (50 HZ OR 60 HZ).

2. NORMAL START (TEMPERATURE ABOVE -25°F)

A. CRANK THE ENGINE BY PLACING THE MASTER SWITCH IN THE START POSITION. DO NOT CRANK FOR CONTINUOUS PERIODS LONGER THAN 15 SECONDS.

B. AT TEMPERATURES BELOW APPROXIMATELY 40°F IT MAY BE NECESSARY TO USE THE AIR HEATER. HOLD THE MASTER SWITCH IN THE PRE-HEAT POSITION FOR 30 SECONDS MAXIMUM PRIOR TO STEP C.

C. HOLD MASTER SWITCH IN START POSITION UNTIL OIL PRESSURE BUILDS UP TO AT LEAST 25 PSI THEN RELEASE TO PRIME & RUN POSITION.

D. ADJUST VOLTAGE AND FREQUENCY TO PROPER VALUES. IF NECESSARY, RESET FAULT INDICATOR LIGHTS.

E. UNDER NORMAL CONDITIONS RUN ENGINE AT NO LOAD FOR 5 MINUTES FOR WARM UP. IF REQUIRED, LOAD CAN BE APPLIED IMMEDIATELY.

F. CLOSE THE AC CIRCUIT INTERRUPTER BY PLACING THE AC CIRCUIT INTERRUPTER SWITCH IN THE CLOSED POSITION.

3. STOPPING THE SET

A. REMOVE LOAD BY PLACING THE AC CIRCUIT INTERRUPTER SWITCH IN OPEN POSITION.

B. ALLOW ENGINE TO OPERATE FOR APPROXIMATELY 5 MINUTES AT NO LOAD.

C. STOP UNIT BY PLACING MASTER SWITCH IN OFF POSITION.

INSTRUCTIONS

4. PARALLEL OPERATION (2 OR MORE LIKE SETS)

A. MAKE CONNECTIONS BETWEEN SETS AND LOAD AS DESCRIBED IN THE OPERATING MANUAL.

B. CONNECT PARALLELING CABLE.

C. START UNITS NO. 1 AND NO. 2 PER STARTING INSTRUCTIONS.

D. ADJUST VOLTAGE AND FREQUENCY TO DESIRED VALUE (MUST BE SAME ON BOTH UNITS).

E. CLOSE AC CIRCUIT INTERRUPTER ON UNIT NO. 1 ONLY.

F. PLACE UNIT-PARALLEL SWITCH ON BOTH SETS IN PARALLEL POSITION.

G. OBSERVE SYNCHRONIZING LIGHTS ON UNIT NO. 2 WHICH SHOULD BE ALTERNATELY GLOWING DARK AND BRIGHT IN UNISON. ADJUST FREQUENCY OF UNIT NO. 2 SLIGHTLY AS NECESSARY TO CAUSE LIGHTS TO SLOWLY GLOW BRIGHT AND DARK IN UNISON.

H. WHEN BOTH LAMPS ARE DARK, CLOSE THE AC CIRCUIT INTERRUPTER ON UNIT NO. 2. (THE UNITS ARE NOW OPERATING IN PARALLEL AND SHOULD APPROXIMATELY DIVIDE KILOWATT LOAD AND CURRENT EQUALLY.)

5. REFER TO APPLICABLE TECHNICAL MANUAL FOR ADDITIONAL INFORMATION ON MAINTENANCE AND TROUBLESHOOTING PROCEDURES.

SERVICE INSTRUCTIONS

FUEL AND OIL

AMBIENT TEMPERATURE	DIESEL FUEL	LUBRICATING OIL
+20°F TO +120°F	VV-F-800 GR DF-2	MIL-L-2104C OE HDO-30
0°F TO +20°F	VV-F-800 GR DF-1	MIL-L-2104C OE HDO-10
-25°F TO 0°F	VV-F-800 GR DF-1	MIL-L-46167
-25°F TO 0°F	VV-F-800 GR DF-A	MIL-L-46167

COOLANT

AMBIENT TEMPERATURE	RADIATOR COOLANT
+40°F TO +120°F	WATER MIL-A-53009
-25°F TO +120°F	WATER MIL-A-46153
-25°F TO +120°F	MIL-A-11755

SYSTEM CAPACITY

FUEL TANK	LUBRICATING OIL		COOLING SYSTEM	
	CRANKCASE	FILTERS	RADIATOR AND OVERFLOW	BLOCK
	FULL	LOW	FILTERS DRAIN TO CRANKCASE	
14 GALLONS	6 QTS.	5 QTS.	-0- QTS.	8 QTS.	3.2 QTS.

NOTE: FOR OPERATION USING JP4, JP5, OR JP8 FUEL REFER TO APPLICABLE OPERATING INSTRUCTION MANUAL.

30554-97-24053

Figure 7. Operating Instructions Plate - MEP-804A/MEP-814A.

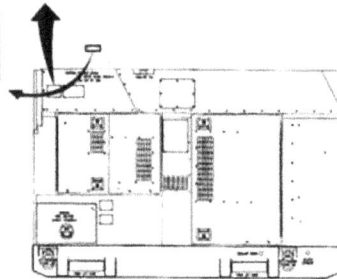

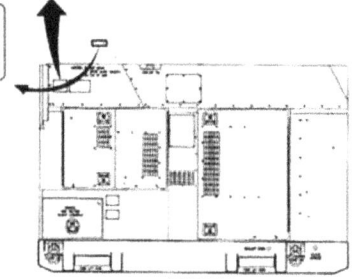

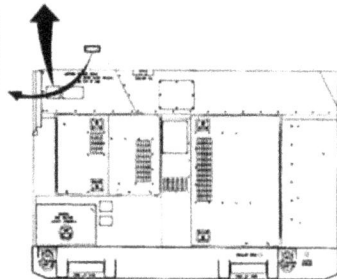

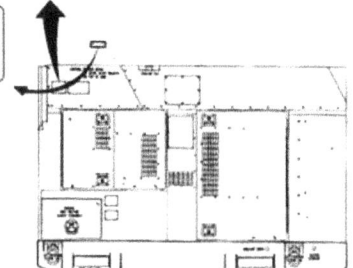

Figure 8. Identification Plates - MEP-804A/MEP-814A.

U.S. DEPARTMENT OF DEFENSE
NATO STANDARD OTAN

GENERATOR SET, DIESEL ENGINE 15KW 50/60 HZ

MODEL MEP-804B   NSN 6115-01-530-1458
SER NO.   REG NO.
TM 9-6115-643-10   NAVFAC 9-6115-643-10
TO 35C2-3-445-21   TM
VOLTS 120/208V 3PH, 240/416V 3PH
AMPS 52. 26   PF 0.8
DRY WT 1785 LB LG 69.7 IN W 35.7 IN HGT 55 IN
DATE MFD   CONTR NO. DAAK01-90-D-0034
WARRANTY FERMONT ESSI   DATE INSP
MFD BY   INSP STAMP

(17V) 30554
(P)MEP-804B
(20S) ZZ20230

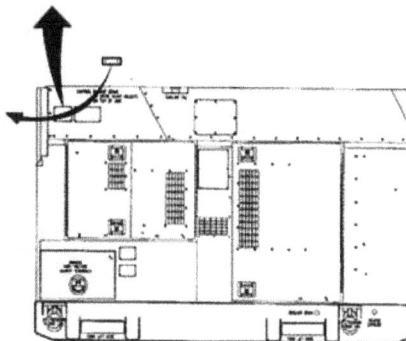

RIGHT SIDE

U.S. DEPARTMENT OF DEFENSE
NATO STANDARD OTAN

GENERATOR SET, DIESEL ENGINE 15KW 50/60 HZ

MODEL MEP-814B   NSN 6115-01-529-9494
SER NO.   REG NO.
TM 9-6115-643-10   NAVFAC 9-6115-643-10
TO 35C2-3-445-21   TM
VOLTS 120/208V 3PH, 240/416V 3PH
AMPS 52. 26   PF 0.8
DRY WT 1915 LB LG 69.7 IN W 35.7 IN HGT 55 IN
DATE MFD   CONTR NO. DAAK01-90-D-0034
WARRANTY FERMONT ESSI   DATE INSP
MFD BY   INSP STAMP

(17V) 30554
(P) MEP-814B
(20S) ZZ20230

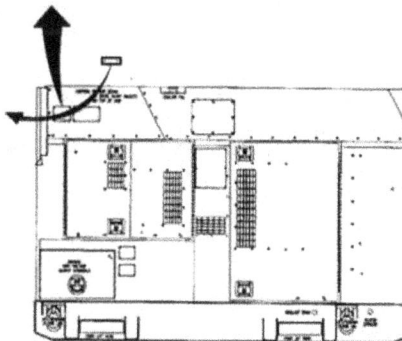

RIGHT SIDE

Figure 9. Identification Plates - MEP-804B/MEP-814B.

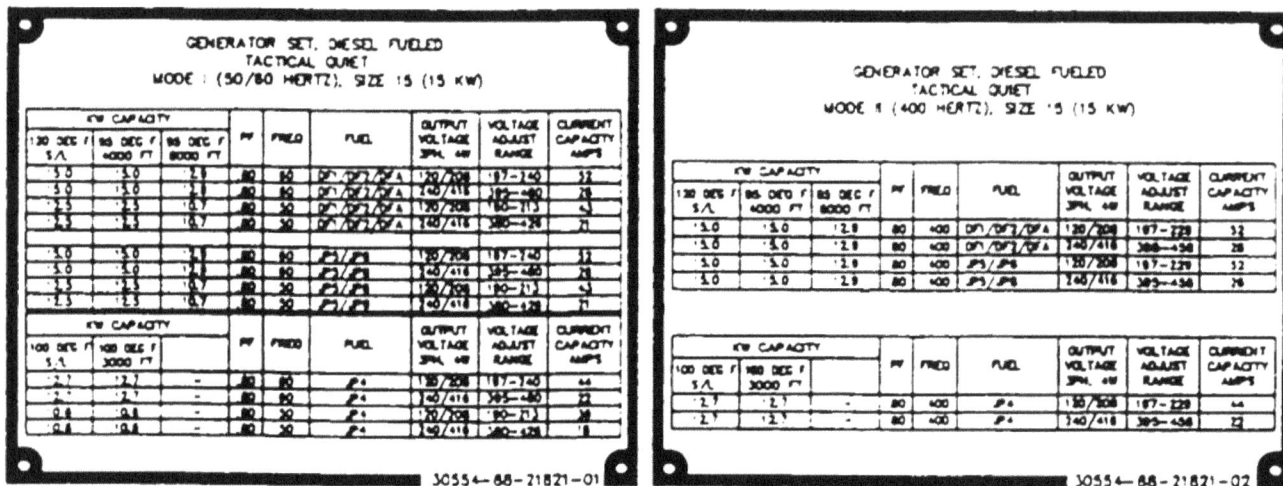

Figure 10. Set Rating Identification Plate.

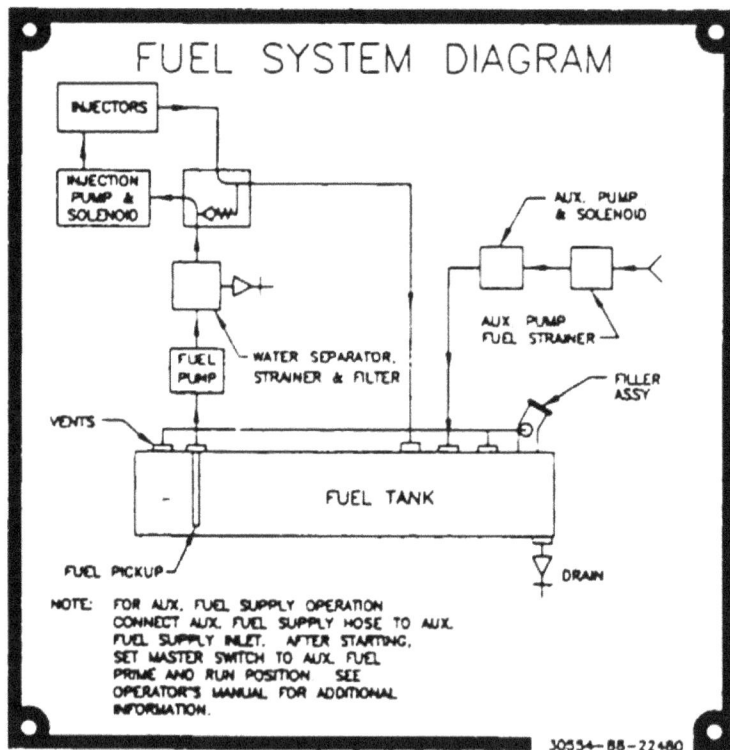

Figure 11. Fuel System Diagram Plate.

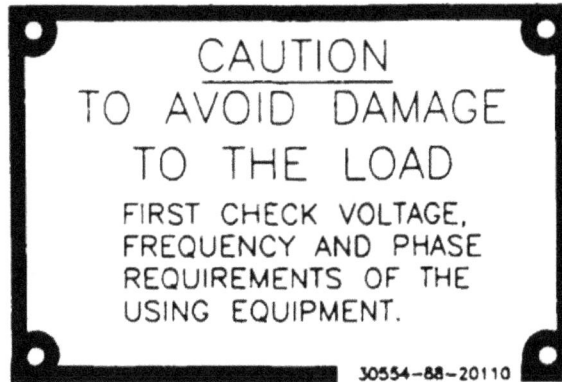

Figure 12. Voltage Connection Caution Plate.

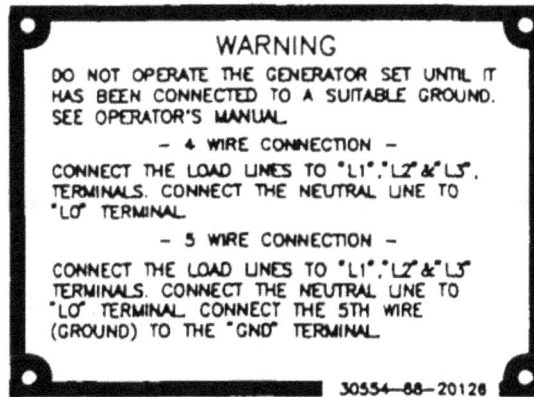

Figure 13. Grounding Stud Plate.

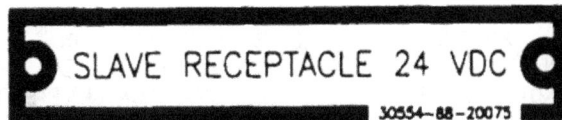

Figure 14. NATO Slave Receptacle Plate.

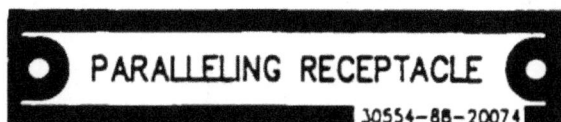

Figure 15. Paralleling Receptacle Plate.

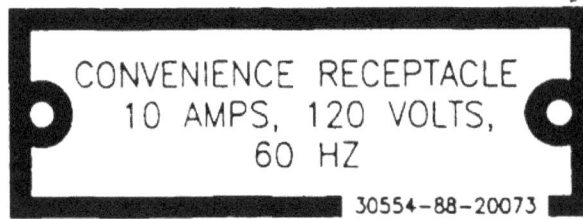

Figure 16. Convenience Receptacle Plate.

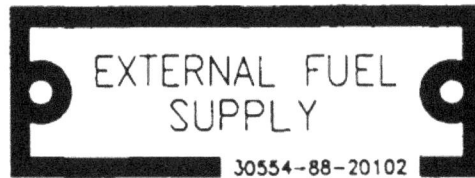

Figure 17. External Fuel Supply Plate.

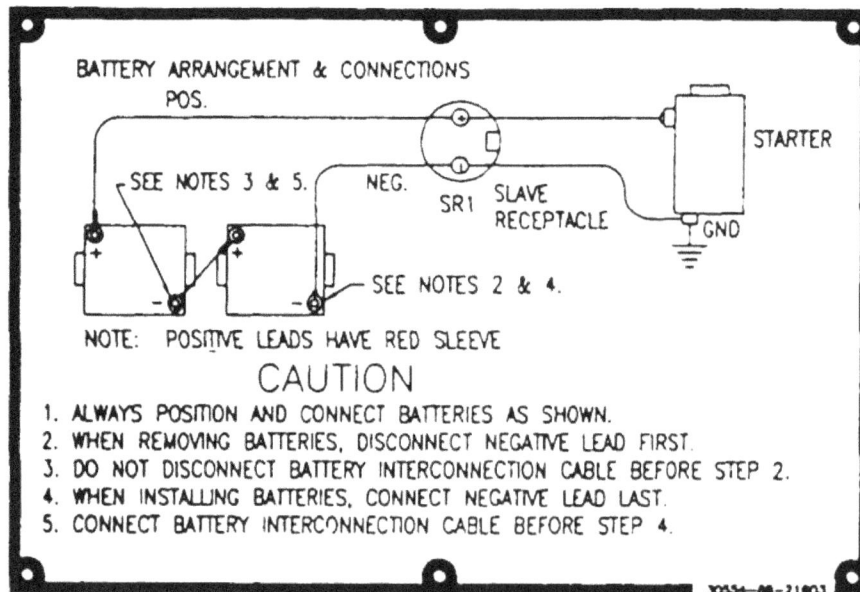

Figure 18. Battery Connection Instruction Plate.

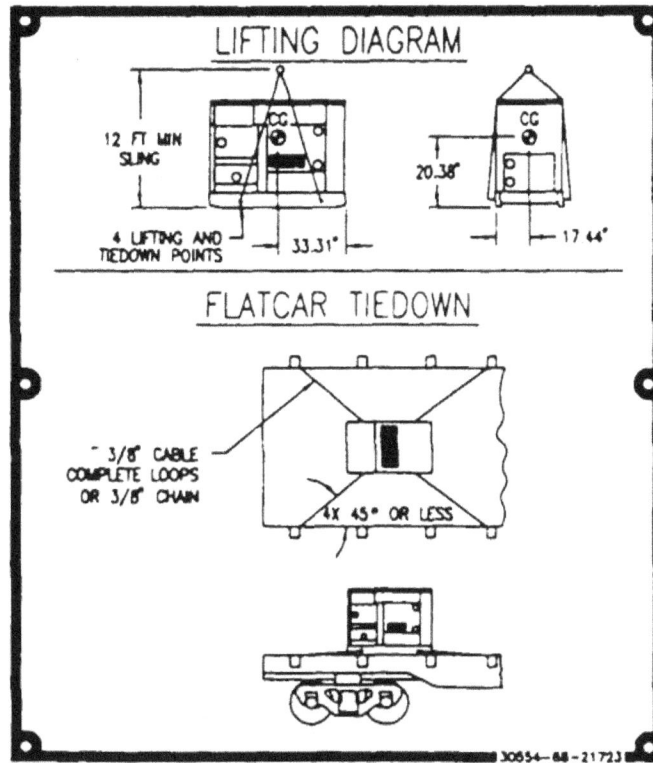

Figure 19. Lifting and Tiedown Diagram Plate.

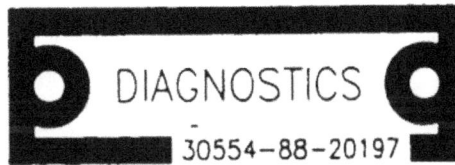

Figure 20. Diagnostics Plate.

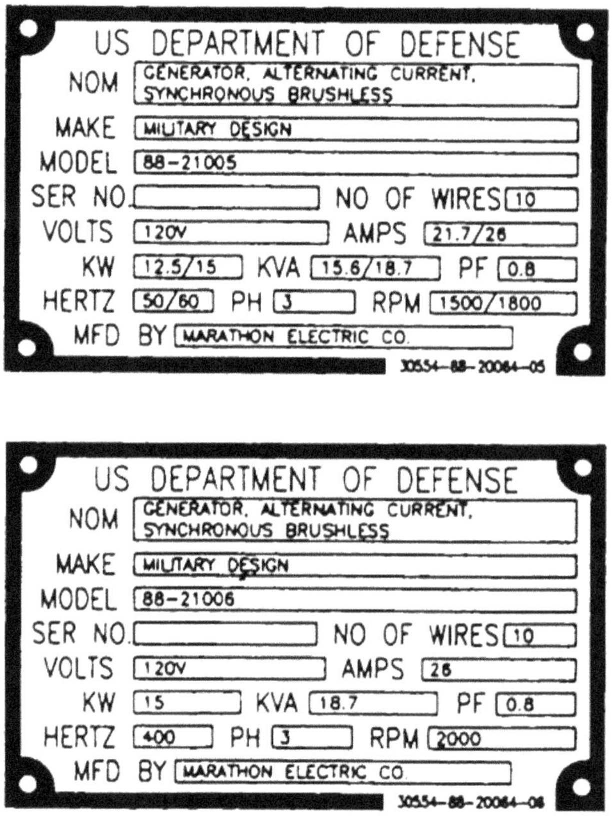

Figure 21. Generator Identification Plate.

PREPARATION FOR MOVEMENT

1. Shut down generator set. Refer to Stopping Procedure. If generator set is operating in parallel, refer to Removal from Parallel Operation.

2. Disconnect load cables.

3. Disconnect paralleling cable, if used, and store in storage box.

4. When using auxiliary fuel line, disconnect line, drain excess fuel from line and store line in storage box.

5. Disconnect ground cable and remove ground rods. Store ground rods in holding clips located inside housing left side generator set. Store cable and couplings in storage box.

6. Secure all generator set access doors and panels.

7. For assembly and preparation for use, refer to Assembly and Preparation for Use.

END OF TASK

END OF WORK PACKAGE

OPERATOR MAINTENANCE

15 kW 50/60 AND 400 Hz SKID MOUNTED, TACTICAL QUIET GENERATOR SET
OPERATION UNDER UNUSUAL CONDITIONS

INITIAL SETUP:

Tools and Special Tools
Generator Mechanical Tool Kit

Personnel Required
One

Equipment Condition
Generator set grounded, off & operational

Materials/Parts
Antifreeze Coolant

References
WP 0005, Starting Procedures

OPERATION IN EXTREME COLD WEATHER BELOW -25 °F (-31 °C)

The generator set operates in ambient temperatures as low as -25°F (-31°C) without special winterization equipment. To ensure satisfactory operation under extreme cold weather the following steps must be taken:

UNUSUAL ENVIRONMENT / WEATHER

Cold Weather Operation

WARNING

All metal jewelry can conduct electricity and become entangled in generator set components. Remove all jewelry when working on generator set. Failure to comply with this warning can cause injury or death to personnel.

WARNING

DO NOT wear loose clothing when performing checks, services and maintenance. Failure to comply with this warning can cause injury or death to personnel.

WARNING

High voltage is produced when the generator set is in operation. DO NOT touch live voltage connections. Never attempt to connect or disconnect load cables or paralleling cables while the generator set is running. Failure to comply with this warning can cause injury or death to personnel.

WARNING

In extreme cold weather, skin can stick to metal. Avoid contacting metal items with bare skin in extreme cold weather. Failure to comply with this warning can cause injury to personnel.

1. Keep generator set and surrounding area as free of ice and snow as practical.
2. Keep fuel tank full to protect against moisture, condensation, and accumulation of water.
3. Ensure that proper grade diesel fuel is used.
4. Keep batteries free from corrosion and in a well charged condition.

END OF TASK

OPERATION IN EXTREME HEAT ABOVE 120 °F (48.8 °C)

1. Check vents and radiator air passages frequently for obstructions.

20.   Check coolant temperature indicator frequently for any indication of overheating.

21.   Allow sufficient space for fuel expansion when filling fuel tank.

22.   Keep generator clean and free of dirt. Clean obstructions from generator intake and outlet

5. Clean external surface of engine when generator set is not operating.

END OF TASK

OPERATION IN DUSTY OR SANDY AREAS

1. If possible, provide a shelter for generator set. Use available natural barriers to shield generator set from blowing dust or sand.

2. Wet down dusty and sandy surface areas around generator set frequently if water is available.

3. Keep all access doors closed, as much as possible, to prevent entry of dust and sand into housing assembly.

4. Wipe dust and sand frequently from the generator set external surface and components. Wash exterior surfaces frequently with clean water when generator set is not operating.

5. Service engine air cleaner assembly frequently to compensate for intake of additional dust or sand.

6. Drain sediment frequently from fuel filter/water separator. When servicing fuel tank be careful to prevent dust or sand from entering fuel tank.

7. Change engine oil and oil filter frequently.

8. Store oil and fuel in dust-free containers.

9. Ensure that generator set ground connections are free of dust and sand and connections are tight before starting the unit.

END OF TASK

OPERATION UNDER RAINY OR HUMID CONDITIONS

CAUTION

Failure to remove waterproof material before operating generator set could result in equipment damage.

1. If possible, provide a shelter for generator set. Cover generator set with canvas or other waterproof material when it is not being operated.

2. Provide adequate drainage to prevent water from accumulating on operation site.

3. Keep all generator set access doors closed, as much as possible, to prevent entry of water into housing assembly.

4. Drain water frequently from fuel filter/water separator.

WARNING

Dangerous voltage exists on live circuits. Always observe precautions and never work alone.
Failure to comply with this warning can cause injury or death to personnel.

5. Remove moisture from generator set components before and after each operating period.

6. Keep fuel tank full to protect against moisture, condensation and accumulation of water.

END OF TASK

OPERATION IN SALT WATER AREAS

CAUTION

Failure to remove waterproof material before operating generator set could result in equipment damage.

1. If possible, provide a shelter for the generator set. Locate generator set so that radiator faces into prevailing winds. Use natural barriers or, if possible, construct a barrier to protect generator set from salt water. Cover generator set with canvas or other waterproof material when it is not being operated.

2. Keep all generator access doors dosed, as much as possible, to prevent entry of salt water into housing assembly.

3. Wash exterior surfaces frequently with dean water when generator set is not operating.

4. Check wiring connections for corrosion and wire insulation for signs of deterioration.

END OF TASK

OPERATION AT HIGH ALTITUDES

The generator set will operate at elevations up to 4000 ft (1219.1 m) above sea level without special adjustment or reduction in load. At elevations greater than 4000 ft (1219.1 m) above sea level, the kilowatt rating is reduced approximately 3.5% for each additional 1000 ft (304.8 m).

END OF TASK

OPERATION WHILE IN CONTAMINATED AREAS

The generator set is capable of being operated by personnel wearing nuclear, biological or chemical (NBC) protective clothing without special tools or supporting equipment. Refer to FM 3-5, NBC Decontamination for information on decontamination procedures. Specific procedures for the generator set are the following:

1. Control panel indicators sealing gasket, rubber sleeves, and rope draw cords at output terminal access ports, control panel door gaskets, access door gaskets, rubber tubing, and belts within the engine compartment, coverings for electrical conduits, external water drain tubing, and retaining cords for slave receptacle covers will absorb and retain chemical agents. Replacement of these items is the recommended method of decontamination.

2. Lubricants, fuel, coolant, or battery fluids may be present on the external surfaces of the generator set or components due to leaks or normal operation. These fluids will absorb NBC agents. The preferred method of decontamination is removal of these fluids using conventional decontamination methods in accordance with FM 3-5.

3. Continued decontamination of external generator set surfaces with supertropical bleach (STB)/ decontamination solution number 2 (DS2) will degrade clear plastic indicator coverings to a point where reading indicators will become impossible. This problem will become more evident for soldiers wearing protective masks. Therefore, the use of STB or DS2 decontamination in these areas should be minimized. Indicators should be decontaminated with warm soapy water.

4. External surfaces of the control panel that are marked with painted or stamped lettering will not withstand repeated decontamination with STB or DS2 without degradation of this lettering. Therefore, the recommended method of decontamination for these areas is with warm soapy water.

5. Areas that will entrap contaminants, making efficient decontamination extremely difficult, include the following:
   - Exposed heads of screws.
   - Areas adjacent to and behind exposed wiring conduits.
   - Hinged areas or access doors.
   - Retaining chains for external receptacle covers.
   - Areas around the tie-down/lifting rings, crevices around access doors, external screens covering ventilation areas, the external oil drain valve, and areas adjacent to the external fuel drain valve.
   - Areas behind knobs and switches on the control panel, externally mounted equipment specification data plates, external receptacle covers, access doors, access door locking mechanisms, recessed wells for access door handles, fuel cap, load terminal board, slave receptacles, and frequency adjustment controls.

NOTE

Replacement of these items, if available, is the preferred method of decontamination. Conventional methods of decontamination should be used on these areas, while stressing the importance of thoroughness and the probability of some degree of continuing contact and vapor hazard.

6. In an NBC contaminated environment, the generator set should be operated with all access doors closed to reduce the effects of contamination.

7. The use of overhead shelters or chemical protective covers is recommended as an additional means of protection against contamination in accordance with FM 3-5. However, if using covers, care should be taken to provide adequate space for air flow and exhaust.

8. For additional NBC information, refer to FM 3-3 and FM 3-4. Other services use applicable publications for NBC.

END OF TASK

USE OF THE CONVENIENCE RECEPTACLE

WARNING

Power is available when the main contactor is open. Avoid accidental contact. Failure to comply with this warning can cause injury or death to personnel.

CAUTION

The maximum power rating for the convenience receptacle is 10 Amps. Continuous operation above 10 Amps can result in damage to the generator set.

1. Start the generator set if it is not operating. Refer to WP 0005, Starting Procedure.

2. Ensure the load does not exceed the maximum rating.

3. Reset the Ground Fault Circuit Interrupter.

4. Plug appropriate connector into convenience receptacle.

END OF TASK

END OF WORK PACKAGE

OPERATOR MAINTENANCE

15 kW 50/60 AND 400 Hz SKID MOUNTED, TACTICAL QUIET GENERATOR SET

EMERGENCY INFORMATION

INITIAL SETUP:

Not Applicable

NATO SLAVE RECEPTACLE START OPERATION

WARNING

All metal jewelry can conduct electricity and become entangled in generator set components. Remove all jewelry when working on generator set. Failure to comply with this warning can cause injury or death to personnel.

WARNING

DO NOT wear loose clothing when performing checks, services and maintenance. Failure to comply with this warning can cause injury or death to personnel.

WARNING

High voltage is produced when the generator set is in operation. DO NOT touch live voltage connections. Never attempt to connect or disconnect load cables or paralleling cables while the generator set is running. Failure to comply with this warning can cause injury or death to personnel.

WARNING

Slave receptacle (NATO connector) is electrically live at all times and is unfused. The Battery Disconnect Switch does not remove power from the slave receptacle. NATO slave receptacle has 24 VDC even when Battery Disconnect Switch is set to OFF. This circuit is only dead when the batteries are fully disconnected. Disconnect the batteries before performing maintenance on the slave receptacle. Failure to comply with this warning can cause injury or death to personnel.

General

The NATO slave receptacle can be used to start the generator set when batteries are discharged.

NATO Slave Emergency Starting Procedure

1. Connect one end of NATO slave cable to fully charged 24 VDC system and other end to discharged generator set's NATO SLAVE RECEPTACLE.

2. Start discharged generator set, refer to WP 0005, Starting Procedure.

3. Remove NATO slave cable after generator set starts.

END OF TASK

EMERGENCY STOPPING

Depressing the EMERGENCY STOP pushbutton will stop the generator set.

NOTE

The generator set cannot be restarted without resetting the EMERGENCY STOP pushbutton and turning the MASTER SWITCH to the OFF position.

END OF TASK

OPERATION USING BATTLE SHORT SWITCH

CAUTION

Continued operation using the BATTLE SHORT switch can result in damage to the generator set.

NOTE

If any emergency situation requires continued operation of the generator set, the BATTLE SHORT switch is used to override all protection devices, and EMERGENCY STOP functions.

NOTE

BATTLE SHORT switch must be OFF to start the generator set.

1. Start generator set, if set is not running. Refer to WP 0005, Starting Procedure.

CAUTION

If the OVERSPEED light on the malfunction indicator panel is illuminated, position the AC circuit interrupter to the OPEN position until indicator goes out on each set.

2. Lift cover on BATTLE SHORT switch and position switch to ON position.

END OF TASK

END OF WORK PACKAGE

CHAPTER 3

OPERATOR TROUBLESHOOTING PROCEDURES

FOR

15 kW 50/60 AND 400 Hz SKID MOUNTED, TACTICAL QUIET GENERATOR SET

CHAPTER 3

OPERATOR TROUBLESHOOTING PROCEDURES

WORK PACKAGE INDEX

OPERATOR MAINTENANCE

15 kW 50/60 AND 400 Hz SKID MOUNTED, TACTICAL QUIET GENERATOR SET
TROUBLESHOOTING INDEX

GENERAL

This work package lists common malfunctions you may find during operation of the generator set. You should perform the tests/inspections and corrective actions in the order listed observing all notes, cautions and warnings.

This manual cannot list all malfunctions that may occur, nor all tests or inspections and corrective actions. If a malfunction is not listed or is not corrected by listed corrective actions, notify your supervisor.

NOTE

Air Force users may perform maintenance only as authorized.

Malfunction/Symptom	Troubleshooting Procedure
ENGINE	
Fails to crank	1.
Cranks but fails to start 2.	
Starts but stops when MASTER SWITCH is released from START position	3.
Stops suddenly	4.
Runs erratically or misfires	5.
Does not develop full power	6.
Knocks	7.
EXHAUST SYSTEM	
Blue or white exhaust smoke	8.
Black exhaust smoke	9.
LUBRICATION SYSTEM	
Low oil pressure	10.
COOLANT SYSTEM	
COOLANT TEMPERATURE indicator	
Indicates engine overheating	11.
ELECTRICAL SYSTEM	
BATTERY CHARGE ammeter	
Shows low or no charge	12.
Shows excessive charging after prolonged operation	13.
AC VOLTMETER (VOLTS AC)	
Indicates low voltage	14.
Indicates correct voltage but frequency meter (HERTZ) is off scale	15.
Fluctuates	16.
FREQUENCY meter (HERTZ)	

TROUBLESHOOTING INDEX - Continued

END OF WORK PACKAGE

OPERATOR MAINTENANCE

15 kW 50/60 AND 400 Hz SKID MOUNTED, TACTICAL QUIET GENERATOR SET
TROUBLESHOOTING PROCEDURES

GENERAL>

This WP lists common malfuntions you may find during operation of the generator set. You should perform the tests/inspections and corrrective actions in order listed.

This manual cannot list all malfunctions that may occur, nor all tests or inspections and corrective actions. If a malfunction is not tested or is not corrected by the list of corrective actions notify your supervisor.

TROUBLESHOOTING PROCEDURES

WARNING

Metal jewelry can conduct electricity and become entangled in generator set components. Remove all metal jewelry when working on generator set. Failure to comply can cause injury or death to personnel.

WARNING

Do not wear loose clothing when performing checks, services and maintenance. Loose clothing may be entangled in generator set components. Failure to comply can cause injury or death to personnel.

WARNING

High voltage is produced when this generator set is in operation. Make sure unit is completely shut down and free of any power source before attempting any repair or maintenance on the unit. Failure to comply can cause injury or death to personnel.

WARNING

High voltage is produced when the generator set is in operation. Never attempt to start the generator set unless it is properly grounded. Failure to comply can cause injury or death to personnel.

SYMPTOM

1. Engine fails to crank

TEST OR INSPECTION

Step 1. Check that DEAD CRANK switch is in the NORMAL position.

CORRECTIVE ACTION

Place switch in NORMAL position.

TEST OR INSPECTION

Step 2. Check that DC CONTROL POWER circuit breaker is energized (in).

CORRECTIVE ACTION

If DC CONTROL POWER circuit breaker is de-energized (out), go to Step 3.

TEST OR INSPECTION

Step 3. Check that Emergency Stop Switch is out.

TEST OR INSPECTION

Step 4. Defect in Engine Starting/Electrical System. Check battery connections.

CORRECTIVE ACTION

If loose or corroded, notify next higher maintenance level.

SYMPTOM

2. Engine cranks but fails to start.

TEST OR INSPECTION

Step 1. Cold ambient temperature.

CORRECTIVE ACTION

If ambient temperature is below 40°F(4°C) turn MASTER SWITCH to PREHEAT for a maximum of 30 seconds prior to cranking engine. Refer to WP 0005, Starting Procedure.

TEST OR INSPECTION

Step 2. Check for dirty air cleaner element.

CORRECTIVE ACTION

Service air cleaner assembly. Refer to WP 0013, Service.

TEST OR INSPECTION

Step 3. Check for dirty fuel filter/water separator.

CORRECTIVE ACTION

Service fuel filter/water separator. Refer to WP 0016, Service. If engine still fails to start, notify next higher maintenance level.

SYMPTOM

3. Engine starts but stops when MASTER SWITCH is released from START position.

TEST OR INSPECTION

Step 1. Check for proper starting procedure.

CORRECTIVE ACTION

Hold MASTER SWITCH in START position until 25 psi (172 kPa) is reached. Refer to WP 0005, Starting Procedures.

TEST OR INSPECTION

Step 2. Check to see if any FAULT INDICATOR lights are lit.

CORRECTIVE ACTION

NO FUEL light is lit. Refer to WP 0014, Service. If any other lights are lit, notify next higher maintenance level.

SYMPTOM

4. Engine stops suddenly.

TEST OR INSPECTION

Step 1. Check to see if any FAULT INDICATOR lights are lit.

CORRECTIVE ACTION

NO FUEL light is lit. Refer to WP 0014, Service. If any other lights are lit, go to Step 2.

TEST OR INSPECTION

Step 2. Check that DC CONTROL POWER circuit breaker is energized (in).

CORRECTIVE ACTION

If DC CONTROL POWER circuit breaker is de-energized (out), notify next higher maintenance level.

SYMPTOM

5. Engine runs erratically or misfires.

TEST OR INSPECTION

Step 1. Check for dirty air cleaner element.

CORRECTIVE ACTION

Service air cleaner assembly. Refer to WP 0013, Service.

TEST OR INSPECTION

Step 2. Check for contaminated fuel.

CORRECTIVE ACTION

Service fuel filter/water separator. Refer to WP 0016, Service.

TEST OR INSPECTION

Step 3. Check for improper type of fuel.

CORRECTIVE ACTION

If improper type of fuel is suspected, refer to WP 0014, Table 1, notify next higher maintenance level.

SYMPTOM

6. Engine does not develop full power.

TEST OR INSPECTION

Step 1. Check for dirty air cleaner element.

CORRECTIVE ACTION

Service air cleaner assembly. Refer to WP 0013, Service.

TEST OR INSPECTION

Step 2. Check for contaminated fuel.

CORRECTIVE ACTION

Service fuel filter/water separator. Refer to WP 0016, Service.

TEST OR INSPECTION

Step 3. Check for restricted exhaust system.

CORRECTIVE ACTION

Make sure exhaust opening is free from obstructions. If no obstructions are found, notify next higher maintenance level.

TEST OR INSPECTION

Step 4. Check for improper type of fuel.

CORRECTIVE ACTION

If improper type of fuel is suspected, refer to WP 0014, Table 1, notify next higher maintenance level.

SYMPTOM

7. Engine knocks.

   TEST OR INSPECTION

   Step 1. Check for low lubrication oil level.

   CORRECTIVE ACTION
   If necessary add oil. Refer to LO 9-6115-643-12.

   TEST OR INSPECTION

   Step 2. Check for loose parts or foreign objects in engine compartment.

   CORRECTIVE ACTION
   If no loose parts or foreign objects are found, go to Step 3.

   TEST OR INSPECTION

   Step 3. Check for improper type of fuel.

   CORRECTIVE ACTION
   If improper type of fuel is suspected, refer to WP 0014, Table 1, notify next higher maintenance level.

SYMPTOM

8. Blue or white exhaust smoke.

   TEST OR INSPECTION

   Check for improper type of fuel.

   CORRECTIVE ACTION
   If improper type of fuel is suspected, refer to WP 0014, Table 1, notify next higher maintenance level.

SYMPTOM

9. Black exhaust smoke.

   TEST OR INSPECTION

   Step 1. Check for improper type of fuel.

   CORRECTIVE ACTION
   If improper type of fuel is suspected, refer to WP 0014, Table 1, notify next higher maintenance level.

   TEST OR INSPECTION

   Step 2. Check for dirty air cleaner element.

   CORRECTIVE ACTION
   Service air cleaner assembly. Refer to WP 0013, Service.

   TEST OR INSPECTION

   Step 3. Check for generator set overload.

   CORRECTIVE ACTION
   Check for generator set overload by checking the ammeter (PERCENT RATED CURRENT) and the kilowattmeter (PERCENT POWER) on the control panel assembly. Refer to WP 0004, Figure 1. If unable to adjust, notify next higher maintenance level.

SYMPTOM

10.   Low oil pressure.

TEST OR INSPECTION

Step 1. Check for low lubrication oil level.
CORRECTIVE ACTION
If necessary add oil. Refer to LO 9-6115-643-12.

TEST OR INSPECTION

Step 2. Check for high coolant temperature, above 200°F(93°C). Refer to WP 0004, Figure 1.
CORRECTIVE ACTION
If coolant temperature is high, go to Step 3.

TEST OR INSPECTION

Step 3. Check coolant level.
CORRECTIVE ACTION
If low, add coolant. Refer to WP 0014, Service. If full, go to Step 4.

TEST OR INSPECTION

Step 4. Check for obstruction in air intake system.
CORRECTIVE ACTION
If obstructions are found, remove debris. If no obstructions are found, go to Step 5.

TEST OR INSPECTION

Step 5. Check for loose fan belt.
CORRECTIVE ACTION
If loose, notify next higher maintenance level.

SYMPTOM

10.   COOLANT TEMPERATURE indicator indicates engine overheating.

TEST OR INSPECTION

Step 1. Check for generator set overload.
CORRECTIVE ACTION
Check for generator set overload by checking the ammeter (PERCENT RATED CURRENT) and the kilowattmeter (PERCENT POWER) on the control panel assembly. Refer to WP 0004, Figure 1. If unable to adjust, notify next higher maintenance level.

TEST OR INSPECTION

Step 2. Check coolant level.
CORRECTIVE ACTION
If low, add coolant. Refer to WP 0014, Service. If full, go to Step 3.

TEST OR INSPECTION

Step 3. Check for low lubrication oil level.

CORRECTIVE ACTION

If necessary add oil. Refer to LO 9-6115-643-12. If full, go to Step 4.

TEST OR INSPECTION

Step 4. Check for obstruction in air intake system.

CORRECTIVE ACTION

If obstructions are found, remove debris. If no obstructions are found, go to Step 5.

TEST OR INSPECTION

Step 5. Check for loose fan belt.

CORRECTIVE ACTION

If loose, notify next higher maintenance level.

SYMPTOM

10.   BATTERY CHARGE ammeter shows low or no charge.

TEST OR INSPECTION

Step 1. Check BATTERY CHARGER FUSE.

CORRECTIVE ACTION

If BATTERY CHARGER FUSE (WP 0004, Figure 1) is blown, notify next higher maintenance level.

TEST OR INSPECTION

Step 2. Check fan belt.

CORRECTIVE ACTION

If loose (WP 0002, Figure 1), notify next higher maintenance level.

TEST OR INSPECTION

Step 3. Check for loose or broken wires.

CORRECTIVE ACTION

Check for loose or broken wires at the back of the battery charging alternator (WP 0002, Figure 1) and BATTERY CHARGE ammeter (WP 0004, Figure 1). If wires are loose or broken, notify next higher maintenance level.

SYMPTOM

10.   BATTERY CHARGE ammeter shows excessive charging after prolonged operation.

TEST OR INSPECTION

Step 1. Check batteries for low electrolyte level.

CORRECTIVE ACTION

If low, refer to WP 0013, Service. If level is correct, go to Step 2.

TEST OR INSPECTION

Step 2. Check battery connections.

CORRECTIVE ACTION

If loose or corroded, notify next higher maintenance level.

SYMPTOM

10.   AC voltmeter (VOLTS AC) indicates low voltage.

TEST OR INSPECTION

Step 1. Check that VM-AM transfer switch position corresponds to readings on the AC voltmeter (VOLTS AC). Refer to WP 0005, Table 1.

CORRECTIVE ACTION

Set VOLTAGE adjust potentiometer.

TEST OR INSPECTION

Step 2. Check for loose or broken wires at back of VM-AM transfer switch, VOLTAGE adjust potentiometer, and AC voltmeter (VOLTS AC).

CORRECTIVE ACTION

If wires are loose or broken, notify next higher maintenance level.

SYMPTOM

10.   AC voltmeter (VOLTS AC) indicates correct voltage, but frequency meter (HERTZ) is off scale.

TEST OR INSPECTION

Step 1. Check FREQUENCY adjust potentiometer

CORRECTIVE ACTION

Set FREQUENCY adjust potentiometer.

TEST OR INSPECTION

Step 2. Check for loose or broken wires at back of FREQUENCY adjust potentiometer.

CORRECTIVE ACTION

If wires are loose or broken, notify next higher maintenance level.

SYMPTOM

10.   AC voltmeter (VOLTS AC) fluctuates.

TEST OR INSPECTION

Check back of AC voltmeter (VOLTS AC) for loose or broken wires.

CORRECTIVE ACTION

If wires are loose or broken, notify next higher maintenance level.

SYMPTOM

10.   Frequency meter (HERTZ) fluctuates.

TEST OR INSPECTION

Check back of frequency meter (HERTZ) for loose or broken wires.

CORRECTIVE ACTION

If wires are loose or broken, notify next higher maintenance level.

SYMPTOM

10.   AC CIRCUIT INTERRUPTER light fails to light when AC CIRCUIT INTERRUPTER switch is closed.

TEST OR INSPECTION

Step 1. Test AC CIRCUIT INTERRUPTER light by depressing.

CORRECTIVE ACTION

If light fails to light, refer to next higher maintenance level.

TEST OR INSPECTION

Step 2. Check load cables for proper connection.

CORRECTIVE ACTION

For proper connection of the load cables, refer to WP 0005, Installation of Load Cables. If correct go to Step 3.

TEST OR INSPECTION

Step 3. Ensure load does not exceed generator rating.

CORRECTIVE ACTION

Ensure load does not exceed generator rating.

SYMPTOM

10.   SYNCHRONIZING LIGHTS fail to light.

TEST OR INSPECTION

Step 1. Check that parallel cable is connected.

CORRECTIVE ACTION

Connect paralleling cable. Refer to WP 0005, Pre-Operation.

TEST OR INSPECTION

Step 2. Check PARALLEL UNIT switch.

CORRECTIVE ACTION

Place PARALLEL UNIT switch in correct position.

SYMPTOM

2. SYNCHRONIZING LIGHTS on generator set No. 2 do not glow bright and dark in unison during parallel operation.

TEST OR INSPECTION

Step 1. Check that load cables are connected properly

CORRECTIVE ACTION

For proper connection of load cables, refer to WP 0005, Installation of Load Cables. If properly connected, go to Step 2.

TEST OR INSPECTION

Step 2. Ensure FREQUENCY SELECT switches are in correct positions.

CORRECTIVE ACTION

Frequency must be the same on both generator sets (MEP-804A/MEP-804B).

SYMPTOM

2. AC CIRCUIT INTERRUPTER light fails to come on generator set No. 2 in parallel operation.

TEST OR INSPECTION

Step 1. Test light by depressing. If light fails to light, notify next higher maintenance level.

CORRECTIVE ACTION

If light comes on, go to Step 2.

TEST OR INSPECTION

Step 2. FREQUENCY adjust potentiometer is not properly adjusted.

CORRECTIVE ACTION
Set FREQUENCY adjust potentiometer.

SYMPTOM

22. No voltage at the Convenience Receptacle.

TEST OR INSPECTION

Step 1. Open control panel and inspect circuit breaker on side of Ground Fault Circuit Interrupter device.

CORRECTIVE ACTION
If tripped, reset device. Check fuse on black wire of Ground Fault Circuit Interrupter for generator sets, contract number DAAK01-88-D-0082.

TEST OR INSPECTION

Step 2. Check reset button for red band.

CORRECTIVE ACTION
If red band is visible, push reset button. If Ground Fault Circuit Interrupter can not be reset, refer to next higher maintenance level.

END OF WORK PACKAGE

CHAPTER 4

OPERATOR MAINTENANCE INSTRUCTIONS

FOR

15 kW 50/60 AND 400 Hz SKID MOUNTED, TACTICAL QUIET GENERATOR SET

CHAPTER 4

OPERATOR MAINTENANCE INSTRUCTIONS

WORK PACKAGE INDEX

OPERATOR MAINTENANCE

15 kW 50/60 AND 400 Hz SKID MOUNTED, TACTICAL QUIET GENERATOR SET

PMCS INTRODUCTION

GENERAL

To ensure that the generator set is ready for operation at all times, it must be inspected so that defects can be discovered and corrected before they result in serious damage or failure. There is no requirement to remove assemblies/equipment prior to performing PMCS

PMCS, BEFORE Operations

Always keep in mind the CAUTIONS and WARNINGS. Perform your Before PMCS.

PMCS, DURING Operations

Always keep in mind the CAUTIONS and WARNINGS. Perform your During PMCS.

PMCS, AFTER Operations

Be sure to perform your After PMCS.

If Your Equipment Fails to Operate

If your equipment does not perform as required, refer to Chapter 3 under Troubleshooting for possible problems. Report any malfunctions or failures on the proper DA Form 2404, or refer to DA PAM 750-8.

PMCS PROCEDURES

NOTE

For general location of the items to be inspected in WP 0011, Table 1, refer to WP 0002, Figure 1, WP 0002, Figure 2, and WP 0004, Figure 1.

Purpose of PMCS Table

Preventive Maintenance Checks and Services (WP 0011, Table 1) list the inspections and care of your equipment required to keep it in good operating condition.

Warnings, Cautions, and Notes

Always observe the *WARNINGS, CAUTIONS,* and *NOTES* appearing in your PMCS table. Warnings and cautions appear before applicable procedures. You must observe *WARNINGS* to prevent serious injury to yourself and others. You must observe *CAUTIONS* to prevent your equipment from being damaged. You must observe *NOTES* to ensure procedures are performed properly.

Explanation of Table Entries

The PMCS table is divided into five columns. Each column is explained in the following paragraphs.

Item No. Column. Numbers in this column are for reference. When completing DA Form 2404 (Equipment Inspection and Maintenance Worksheet), include the item number for the check/service indicating a fault. Item numbers also appear in the order that you must do checks and services for the intervals listed.

Interval Column. This column tells you when you must do the procedure described in the procedure column. "BEFORE" procedures must be done before you operate the equipment for its intended mission. "DURING" procedures must be done during the time you are operating the equipment for its intended mission. "AFTER" procedures must be done immediately after you have operated the equipment. Perform "WEEKLY" procedures at the listed interval.

Location, Item to Check/Service Column. This column lists the location and the item to be checked or serviced. The item location is underlined.

Procedure Column. This column gives the procedure for checking or servicing the item listed in the location, item to check/service column. You must perform the procedure to know if the power unit or power plant is ready or available for its intended mission or operation. You must do the procedure at the time stated in the interval column.

Equipment Not Ready/Available if: Column. Information in this column tells you what faults will keep your equipment from being capable of performing its primary mission. If you perform checks or services that show faults listed in this column, do not operate the equipment.

Other Table Entries

Be sure to observe all special information and notes that appear in your table.

Special Instructions

Preventive maintenance is not limited to performing the checks and services listed in the PMCS Table. Covering unused receptacles, stowing unused accessories and performing other routine procedures such as equipment inventory, cleaning components, and touch-up painting are not listed in the table. These are things you should do any time you see that they need to be done. If a routine check is listed in the PMCS Table, it is because experience has shown that problems may occur with this item. Take along tools and cleaning cloths needed to perform the required checks and services. Use the information in the following paragraphs to help you identify problems at any time and to help identify potential problems before and during checks and services.

WARNING

Metal jewelry can conduct electricity and become entangled in generator set components. Remove all metal jewelry when working on generator set. Failure to comply can cause injury or death to personnel.

WARNING

Do not wear loose clothing when performing checks, services and maintenance. Loose clothing may be entangled in generator set components. Failure to comply can cause injury or death to personnel.

WARNING

High voltage is produced when this generator set is in operation. Make sure unit is completely shut down and free of any power source before attempting any repair or maintenance on the unit. Failure to comply can cause injury or death to personnel.

WARNING

Solvent used to clean parts is potentially dangerous to personnel and property. Clean parts in a well-ventilated area. Avoid inhalation of solvent fumes. Wear goggles and rubber gloves to protect eyes and skin. Wash exposed skin thoroughly. Do not smoke or use near open flame or excessive heat. Failure to comply with this warning can cause injury to personnel, and damage to the equipment.

CAUTION

Keep cleaning solvents, fuels and lubricants away from rubber or soft plastic parts. They will deteriorate material.

1. Keep the generator set clean. Dirt, grease, and oil get in the way and may cover up a serious problem. Use cleaning solvent to clean metal surfaces.
2. Use soap and water to clean rubber or plastic parts and material.
3. Check all bolts, nuts, and screws to make sure they are not loose, missing, bent, or broken. Do not try to check them with a tool, but look for chipped paint, bare metal, or rust around bolt heads. If you find one loose,

report it to the next-higher level of maintenance.

4. Inspect welds for loose or chipped paint, rust, or gaps where parts are welded together. If a broken weld is found, report it to the next-higher level of maintenance.

5. Inspect electrical wires, connectors, terminals, and receptacles for cracked or broken insulation, bare wires, and loose or broken connectors. Tighten loose connectors. Examine terminals and receptacles for serviceability. If deficiencies are found, report them to the next-higher level of maintenance.

6. Inspect hoses and fluid lines. Look for wear, damage, and leaks. Make sure that clamps and fittings are tight. Wet spots and stains around a fitting or connector can mean a leak. If a leak comes from a loose connector, or if something is broken or worn out, report it to the next-higher level of maintenance.

Leakage Definitions

You must know how fluid leakage affects the status of your equipment. The following are definitions of the types/classes of leakage you need to know to be able to determine the status of your equipment. Learn and be familiar with them. When in doubt, notify your supervisor.

Table 1. Leakage Definitions.

Leakage Class Leakage Definition

Class I Seepage of fluid (as indicated by wetness or discoloration) not great enough to form drops.

Class II Leakage of fluid great enough to form drops, but not enough to cause drops to drip from the item being checked/inspected.

Class III Leakage of fluid (other than fuel) greater than three drops per minute that fall from the item being inspected.

Operation of Generator Set with Minor Leaks

CAUTION

Equipment operation is allowable with minor leakage (Class I or II) of any fluid except fuel. Fluid capacity must be considered before deciding to continue operation of the equipment with minor leaks. When operating with Class I or II leaks, fluid level must be checked more often than required by the PMCS table. Parts without fluid will stop working and/or cause equipment damage.

a. Consider the equipment's capacity for the fluid that is leaking. If the capacity is small, the fluid level may soon become too low for continued operation. If in doubt, notify your supervisor.

b. Check the fluid level more often than required in the PMCS Table. Add fluid as needed.

c. All leaks should be reported to the next higher level of maintenance.

Corrosion Prevention and Control (CPC)

CPC of Army material is of continuing concern. It is important that any corrosion problems with the equipment be reported so that the problem can be corrected and improvements can be made to prevent the problem in future items. Although corrosion is typically associated with rusting of metals, it can also include deterioration of other materials, such as rubber and plastic. Unusual cracking, softening, swelling, or breaking of these materials may be a corrosion problem. If a corrosion problem is identified, it can be reported using Standard Form 368, Product Quality Deficiency Report. Use of key words such as "corrosion," "rust," "deterioration," or "cracking" will ensure that the information is identified as a CPC problem. The form should be submitted to the address specified in DA PAM 750-8.

Removal of Assemblies/Equipment to Perform PMCS

There is no requirement to remove assemblies/equipment prior to performing the PMCS.

Winterization Kit

See Chapter 4 for PMCS Procedures.

END OF WORK PACKAGE

OPERATOR MAINTENANCE

15 kW 50/60 AND 400 Hz SKID MOUNTED, TACTICAL QUIET GENERATOR SET
PMCS, INCLUDING LUBRICATION INSTRUCTIONS

INITIAL SETUP:

Tools and Special Tools
Generator Mechanical Tool Kit

Personnel Required
One

Equipment Condition
Generator set grounded, off & operational

Materials/Parts
Expendable Durable/Items List

References
LO 9-6115-643-10

LUBRICATION ORDER

Refer to LO 9-6115-643-12 for lubrication information.

Table 1. Preventive Maintenance Checks and Services for Model MEP-804A, MEP-804B, MEP-814A and MEP-814B.

ITEM NO.	INTERVAL	ITEM TO BE CHECKED OR SERVICED	PROCEDURE	EQUIPMENT NOT READY/AVAILABLE IF:
		GENERATORSET EXTERIOR		

WARNING

In extreme cold weather, skin can stick to metal. Avoid contacting metal items with bare skin inextreme cold weather. Failure to comply can cause injury or death to personnel.

NOTE

The generator set can be operated continuously at any load from no load up to and including ratedload. However, at light loads (less than 25% of set rating), an oily residue (unburned fuel oil) mayoccasionally be noticed In the exhaust system outlet and around connection joints in the exhaustsystem. This residue is caused by the inability of the fuel injection system to consistently meter thesmall amount of fuel required to operate at these low load levels and is not a defect in the fuelsystem. The oily residue could affect engine performance and create a cosmetic problem on andaround the generator set. Operation at rated load will burn off this oily residue. The length of timerequired at rated load depends on the amount of residue. The muffler may also need to be removedand cleaned if excessive build up occurs. This oily residue can be prevented by increasing the elec-trical load on the set.

NOTE

If the equipment must be kept in continuous operation, check and service only those Items that canbe checked and serviced without disrupting operations. Complete all checks and services whenequipment is shut down.

| 1 | Before | HOUSING | Check door panels, hinges, and latches for damage, loose, orcorroded items. | Cannot secure door. |

Table 1. Preventive Maintenance Checks and Services for Model MEP-804A, MEP-804B, MEP-814A and MEP-814B. - Continued

ITEM NO.	INTERVAL	ITEM TO BE CHECKED OR SERVICED	PROCEDURE	EQUIPMENT NOT READY/AVAILABLE IF:
			Inspect air intake and exhaust grillsfor debris.	
2	Before	IDENTIFICATION PLATES	Check to ensure identification plate-sare secure.	
3	Before	SKID BASE	Inspect skid base for cracks and/orcor-rosion.	Skid base is cracked orshows signs of structur-aldamage.
4	Before	ACOUSTICAL MATER-IALS	Ensure that acoustical materials are-free of damage and not missing.	

ENGINEASSEMBLY

WARNING

Operating the generator set exposes personnel to a high noise level. Hearing protection must beworn when operating or working near the generator set when the generator set is running. Failure tocomply can cause hearing damage to personnel.

WARNING

Fuels used in the generator set are combustible. Do not smoke or use open flames when performingmaintenance. Failure to comply can result in flames and possible explosion and can cause injury ordeath to personnel and damage to the generator set.

5 Before ENGINE ASSEMBLY Inspect for loose, damaged, or

Any loose, damaged, ormissing hardware.
missing hardware.

6	Before	FUEL SYSTEM	Inspect for leaks, damage, loose, orm-issing hardware.	Any fuel leaks, damage,-loose or missing parts.
7	Before	FUEL FILTER/WATER SEPARATOR	Inspect for leaks, cracks, dam-age,proper mounting, loose or missing parts.	Any fuel leaks.
			Drain water from fuel filter/watersep-arator.	Water not drained.
8	Before	LUBRICATION SYSTEM		
Pull sert Pull	dipstick from dip-stick into dipstick from	oil dipstick tube and take oil dipstick tube (dipstick oil dipstick tube and take	**NOTE** reading. If recheck is desired: Wipe dipstick clean. Rein- must remain in oil dipstick tube for 5 seconds minimum). reading. Inspect for leaks, damage, loose ormissing parts. Class III leaks, damage,loose or missing parts. Inspect oil level. Oil level is low. Inspect for contamination. Oil shows signs of contamination.	
		COOLINGSYSTEM		

Table 1. Preventive Maintenance Checks and Services for Model MEP-804A, MEP-804B, MEP-814A and MEP-814B. - Continued

ITEM NO.	INTERVAL	ITEM TO BE CHECKED OR SERVICED	PROCEDURE	EQUIPMENT NOT READY/AVAILABLE IF:
WARNING				
Cooling system operates at high temperature and pressure. Contact with high pressure steam and/or liquids can result in burns and scalding. Shut down generator set, and allow system to cool before performing checks, services and maintenance. Failure to comply can cause injury or death to personnel.				
9	Before	RADIATOR	Inspect for leaks, damage, loose or missing parts.	Class III leaks or missing radiator cap.
10	Before	HOSES	Inspect for leaks, cracks, or missing parts.	Class III leaks or missing clamps or hoses.
11	Before	COOLING FAN	Inspect for obstruction, damage, or looseness.	Damaged or loose.
			Inspect for unusual noise in fan area.	Unusual noise from area.
11.1	Before	WATER PUMP	Inspect for leaks.	Class III leaks or unusual noise from area.
12	Before	FAN BELT	Inspect for cracks, fraying, or looseness.	Broken or missing belt.
13	Before	OVERFLOW BOTTLE	Inspect for proper mounting, leaks, or missing hardware.	Class III leaks or missing hardware.
EXHAUST/INTAKE SYSTEM				
WARNING				
Exhaust discharge contains deadly gases including carbon monoxide. Do not operate generator set in an enclosed area unless exhaust discharge is properly vented outside. Failure to comply can cause injury or death to personnel.				
14	Before	EXHAUST SYSTEM	Inspect for leaks, corrosion, and missing parts.	Leaks, damaged, or missing parts.
15	Before	AIR CLEANER ASSEMBLY	Inspect for loose, damaged, or missing parts.	Loose or missing parts.
			Inspect restriction indicator for clogged air cleaner element.	Clogged air cleaner element.
15.1	Before	CRANKCASE VENTILATION FILTER (MEP-804B/MEP-814B)	Inspect for loose, damaged, or missing parts.	Loose or missing parts.
			Remove filter element and inspect for oil saturation or damaged filter element.	Clogged or damaged.
		GROUNDING ROD ASSEMBLY		

Table 1. Preventive Maintenance Checks and Services for Model MEP-804A, MEP-804B, MEP-814A and MEP-814B. - Continued

ITEM NO.	INTERVAL	ITEM TO BE CHECKED OR SERVICED	PROCEDURE	EQUIPMENT NOT READY/AVAILABLE IF:
			WARNING	
		High voltage is produced when the generator set is in operation. Never attempt to start the generatorset unless it is properly grounded. Failure to comply can cause injury or death to personnel.		
		WARNING		
		Ensure nuts on ground terminals are properly secured creating a good ground. Failure to comply withthis warning can cause injury or death to personnel.		
16	Before	GROUND ROD CABLE AND CONNECTIONS	Inspect for damage, corrosion, andloose connections.	Damaged, corroded, orloose connections.
		ELECTRICALSYSTEM		
		WARNING		
		Batteries give off a flammable gas. Do not smoke or use open flame when performing maintenance. Failure to comply can cause injury or death to personnel and equipment damage due to flames andexplosion.		
		WARNING		
		Battery acid can cause burns to skin and cause eye injury. Wear safety goggles and chemical glovesand avoid acid splash while working on the batteries. Failure to comply may cause injury or death topersonnel.		
		WARNING		
		Dangerous voltage exists on live circuits. Always observe precautions and never work alone. Failureto comply with this warning can cause injury or death to personnel.		
17	Before	BATTERIES	Inspect electrolyte level.	Electrolyte is below battery plates.
18	Before	BATTERY CABLES	Inspect for corrosion, damage, looseconnections, or missing parts.	Damaged, loose, ormissing parts.
19	Before	OUTPUT BOX ASSEMBLY	Inspect cables for damage or looseconnections.	Damaged, loose, ormissing parts.
			Inspect output terminals for damageor missing hardware.	Damaged or missinghardware.
20	Before	CONTROLBOX ASSEMBLY CONTROLS AND INDICATORS	Inspect for damage or missing parts.	Damaged or missingparts.

WARNING

High voltage is produced when this generator set is in operation. Make sure unit is completely shut down and free of any power source before attempting any repair or maintenance on the unit. Failure to comply can cause injury or death to personnel.

Table 1. Preventive Maintenance Checks and Services for Model MEP-804A, MEP-804B, MEP-814A and MEP-814B. - Continued

ITEM NO.	INTERVAL	ITEM TO BE CHECKED OR SERVICED	PROCEDURE	EQUIPMENT NOT READY/AVAILABLE IF:
21	Before	CONTROL BOX HARNESS	Inspect for damage and looseness.	Damaged or loose.

GENERATORSET
EXTERIOR

WARNING

Operating the generator set exposes personnel to a high noise level. Hearing protection must beworn when operating or working near the generator set when the generator set is running. Failure tocomply can cause hearing damage to personnel.

WARNING

Fuels used in the generator set are combustible. Do not smoke or use open flames when performingmaintenance. Failure to comply can result in flames and possible explosion and can cause injury ordeath to personnel and damage to the generator set.

WARNING

Top housing panels can get very hot. When performing DURING PMCS, wear gloves and additionalprotective clothing as required. Failure to comply can result in severe burns to personnel.

NOTE

If the equipment must be kept in continuous operation, check and service only those items that canbe checked and serviced without disrupting operations. Complete all checks and services whenequipment is shut down.

ITEM NO.	INTERVAL	ITEM TO BE CHECKED OR SERVICED	PROCEDURE	EQUIPMENT NOT READY/AVAILABLE IF:
22	During	HOUSING	Check door panels, hinges, and latches for damage, loose, orcorroded items.	Cannot secure door.
23	During	~~ENGINEASSEMBLY~~ ENGINE ASSEMBLY	Inspect for loose, damaged, ormissing hardware.	Any loose, damaged, ormissing hardware.
24	During	FUEL SYSTEM	Inspect for leaks, damage, loose, ormissing hardware.	Any fuel leaks, damage,-loose or missing parts.
25	During	LUBRICATION SYSTEM		

Table 1. Preventive Maintenance Checks and Services for Model MEP-804A, MEP-804B, MEP-814A and MEP-814B. - Continued

ITEM NO.	INTERVAL	ITEM TO BE CHECKED OR SERVICED	PROCEDURE	EQUIPMENT NOT READY/AVAILABLE IF:
			NOTE	

Pull dipstick from oil dipstick tube and take reading. If recheck is desired:
- Wipe dipstick clean.
- Reinsert dipstick into oil dipstick tube (dipstick must remain in oil dipstick tube for 5 seconds minimum).
- Pull dipstick from oil dipstick tube and take reading.

Inspect for leaks, damage, loose or missing parts. Class III leaks, damage, loose or missing parts.

Inspect oil level. Oil level is low.

Inspect for contamination. Oil shows signs of contamination.

COOLING SYSTEM

26 During COOLING FAN Inspect for obstruction, damage, or looseness. Damaged or loose.

Inspect for unusual noise in fan area. Unusual noise from area.

| 27 | During | OVERFLOW BOTTLE | Inspect for proper mounting, leaks, or missing hardware. | Class III leaks or missing hardware. |
| 28 | During | **GROUNDING ROD ASSEMBLY** GROUND ROD CABLE AND CONNECTIONS | Inspect for damage, corrosion, and loose connections. | Damaged, corroded, or loose connections. |

CONTROL BOX ASSEMBLY

WARNING

High voltage is produced when the generator set is in operation. Never attempt to connect or disconnect load cables while the generator set is running. Failure to comply can cause injury or death to personnel.

29 During CONTROLS AND INDICATORS   Inspect indicators are operating properly. Indicators are not operating properly.

NOTE

If the equipment must be kept in service continuous operation, check only those items that can be checked and serviced without disrupting operations. Complete all checks and services when equipment is shut down.

GENERATOR SET

Table 1. Preventive Maintenance Checks and Services for Model MEP-804A, MEP-804B, MEP-814A and MEP-814B. - Continued

ITEM NO.	INTERVAL	ITEM TO BE CHECKED OR SERVICED	PROCEDURE	EQUIPMENT NOT READY/AVAILABLE IF:
		WARNING		

Exhaust system can get very hot. Shut down generator set, and allow system to cool beforeperforming checks, services and maintenance. Failure to comply can cause severe burns and injuryto personnel.

WARNING

Top housing panels can get very hot. Allow panels to cool down before performing maintenance. Failure to comply can result in severe burns to personnel.

30 After HOUSING Check door panels, hinges, and latches for damage, loose, orcorroded items. — Cannot secure door.

| 31 | After | IDENTIFICATION PLATES | Check to ensure identification plate-sare secure. | |
| 32 | After | SKID BASE | Inspect skid base for cracks and/ orcorrosion. | Skid base is cracked orshows signs of structur-aldamage. |

WARNING

Fuels used in the generator set are combustible. Do not smoke or use open flames when performingmaintenance. Failure to comply can result in flames and possible explosion and can cause injury ordeath to personnel and damage to the generator set.

WARNING

All fuel is combustible and toxic to eyes, skin, and respiratory tract. Skin and eye protection arerequired when working in contact with diesel fuel. Avoid repeated or prolonged contact. Provideadequate ventilation. Operators are to wash skin exposed to fuel and change fuel soaked clothingpromptly. Failure to comply can cause serious injury to personnel.

33 After ENGINE ASSEMBLY Inspect for loose, damaged, or missing hardware. — Loose, damaged, ormissing hardware.

		FUELSYSTEM		
34	After	FUEL SYSTEM	Inspect for leaks, damage, loose, orm-issing hardware.	Any fuel leaks, damage,-loose or missing parts.
35	After	FUEL FILTER/ WATER SEPARATOR	Inspect for leaks, cracks, dam-age,proper mounting, loose or miss-ingparts.	Any fuel leaks.
36	After	LUBRICATION SYSTEM		

Table 1. Preventive Maintenance Checks and Services for Model MEP-804A, MEP-804B, MEP-814A and MEP-814B. - Continued

ITEM NO.	INTERVAL	ITEM TO BE CHECKED OR SERVICED	PROCEDURE	EQUIPMENT NOT READY/AVAILABLE IF:

NOTE

Pull dipstick from oil dipstick tube and take reading. If recheck is desired:
- Wipe dipstick clean.
- Reinsert dipstick into oil dipstick tube (dipstick must remain in oil dipstick tube for 5 seconds minimum).
- Pull dipstick from oil dipstick tube and take reading.

			Inspect for leaks, damage, loose or missing parts.	Class III leaks, damage, loose or missing parts.
			Drain water.	Water not drained.
			Inspect oil level.	Oil level is low.
			Inspect for contamination.	Oil shows signs of contamination.
		COOLING SYSTEM		

WARNING

Cooling system operates at high temperature and pressure. Contact with high pressure steam and/or liquids can result in burns and scalding. Shut down generator set, and allow system to cool before performing checks, services and maintenance. Failure to comply can cause injury or death to personnel.

37	After	RADIATOR	Inspect for leaks, damage, loose or	Class III leaks or missing
38	After	HOSES	Inspect for leaks, cracks, or missing parts.	Class III leaks or missing clamps or hoses.
39	After	FAN BELT	Inspect for cracks, fraying, or looseness.	Broken or missing belt.
40	After	**CONTROLBOX ASSEMBLY** CONTROLS AND INDICATORS	Inspect for damaged or missing parts.	Damaged or missing parts.

Mandatory Replacement Parts List

There are no replacement parts required for these PMCS procedures.

END OF WORK PACKAGE

OPERATOR MAINTENANCE

15 kW 50/60 AND 400 Hz SKID MOUNTED, TACTICAL QUIET GENERATOR SET

BATTERIES: INSPECTION, SERVICE

INITIAL SETUP:

Tools and Special Tools
Generator Mechanical Tool Kit

Materials/Parts
Safety Goggles
Chemical Gloves
Distilled Water

Personnel Required
One

References
WP 0005, Starting/Stopping Procedures

Equipment Condition
Generator set ground, off & operational

INTRODUCTION

WP 0012 through WP 0018 contain operator maintenance procedures. Deficiencies noted during inspection which are beyond the maintenance scope of the operator shall be reported to next higher maintenance

WARNING

Metal jewelry can conduct electricity and become entangled in generator set components. Remove all metal jewelry when working on generator set. Failure to comply can cause injury or death to personnel.

WARNING

Do not wear loose clothing when performing checks, services and maintenance. Loose clothing may be entangled in generator set components. Failure to comply can cause injury or death to personnel.

WARNING

Battery acid can cause burns to skin and cause eye injury. Wear safety goggles and chemical gloves and avoid acid splash while working on the batteries. Failure to comply may cause injury or death to personnel.

WARNING

Batteries give off a flammable gas. Do not smoke or use open flame when performing maintenance. Failure to comply can cause injury or death to personnel and equipment damage due to flames and explosion.

INSPECTION

1. Shut down generator set. Refer to WP 0005, Stopping Procedure.

2. Open battery access door.

3. Inspect for damaged battery case, corrosion, or damaged and loose connections on terminal cable, and damaged or missing battery caps.

WARNING

Batteries give off a flammable gas. Do not smoke or use open flame when performing maintenance. Failure to comply can cause injury or death to personnel and equipment damage due to flames and explosion.

4. Remove battery caps.

CAUTION

Electrolyte level must cover battery plates in all cells. Failure to observe this caution can cause damage to the battery.

NOTE

Electrolyte level should be at bottom of each cap cylinder.

5. Inspect electrolyte level.

6. Perform service procedures if required.

7. Install battery caps.

8. Close battery access door.

END OF TASK

SERVICING

1. Shut down generator set. Refer to WP 0005, Stopping Procedure.

2. Open battery access door.

WARNING

Metal jewelry can conduct electricity and become entangled in generator set components. Remove all metal jewelry when working on generator set. Failure to comply can cause injury or death to personnel.

WARNING

Do not wear loose clothing when performing checks, services and maintenance. Loose clothing may be entangled in generator set components. Failure to comply can cause injury or death to personnel.

WARNING

Batteries give off a flammable gas. Do not smoke or use open flame when performing maintenance. Failure to comply can cause injury or death to personnel and equipment damage due to flames and explosion.

3. Remove battery caps.

NOTE

Electrolyte level should be at bottom of each cap cylinder.

4. Add distilled water to each battery cell as required.

5. Replace battery caps.

6. Close battery access door.

7. If necessary contact next higher level of maintenance to clean or replace batteries or battery terminals.

END OF TASK

END OF WORK PACKAGE

OPERATOR MAINTENANCE

15 kW 50/60 AND 400 Hz SKID MOUNTED, TACTICAL QUIET GENERATOR SET
AIR CLEANER ASSEMBLY: INSPECTION, SERVICE

INITIAL SETUP:

Tools and Special Tools
Generator Mechanical Tool Kit

Personnel Required
One

Equipment Condition
Generator set grounded, off & operational

Materials/Parts
Air cleaner element
Clean lint-free cloth (WP 0022)

References
WP 0005, Stopping Procedures

INSPECTION

1. Shut down generator set. Refer to WP 0005, Stopping Procedure.

2. Open air cleaner access door (rear of generator set).

3. Open left side engine compartment access door.

4. Inspect air cleaner housing (5) for dents, corrosion, missing hardware and other damage.

5. Inspect restriction indicator (6) for indication of a clogged air cleaner element (4).

6. Close air cleaner access door.

END OF TASK

SERVICING

1. Shut down generator set. Refer to WP 0005, Stopping Procedure.

2. Open air cleaner access door (rear of generator set).

3. Loosen retaining clamp (Figure 1, Item 1) and remove end cap (2) on air cleaner housing (5).

4. Remove wing nut (3) and air cleaner element (4). If fouled, discard air cleaner element.

5. Inspect inside of air cleaner housing (5) for debris. Wipe housing interior with clean lint-free cloth (WP 0022, Item 7).

6. Install air cleaner element (4), wing nut (3), end cap (2) and hand tighten retaining clamp (1).

7. Close air cleaner access door.

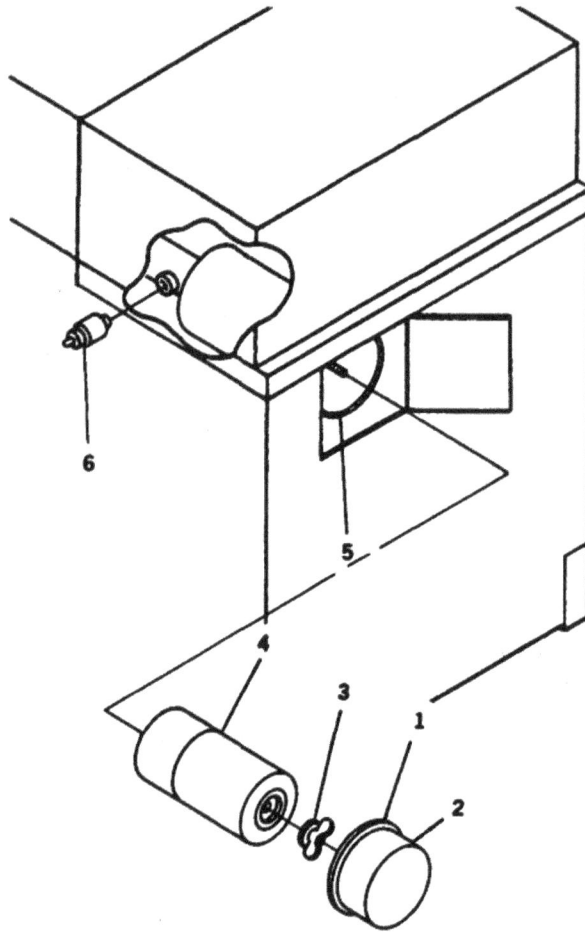

Figure 1. Air Cleaner Element Replacement.

END OF TASK

END OF WORK PACKAGE

OPERATOR MAINTENANCE

15 kW 50/60 AND 400 Hz SKID MOUNTED, TACTICAL QUIET GENERATOR SET

COOLING SYSTEM: INSPECTION, SERVICE

INITIAL SETUP:

Tools and Special Tools
Generator Mechanical Tool Kit

Personnel Required
One

Equipment Condition
Grounded off & operational

Materials/Parts
Coolant (MIL-A-53009 A) & Antifreeze (A-A-52624 A

References
WP 0005, Stopping Procedures

INSPECTION

WARNING

Metal jewelry can conduct electricity and become entangled in generator set components. Remove all metal jewelry when working on generator set. Failure to comply can cause injury or death to personnel.

WARNING

Do not wear loose clothing when performing checks, services and maintenance. Loose clothing may be entangled in generator set components. Failure to comply can cause injury or death to personnel.

1. Shut down generator set. Refer to WP 0005, Stopping Procedure.

2. Open both engine access doors.

WARNING

Cooling system operates at high temperature and pressure. Contact with high pressure steam and/ or liquids can result in burns and scalding. Shut down generator set, and allow system to cool before performing checks, services and maintenance. Failure to comply can cause injury or death to personnel.

3. Check radiator for dirt, leaves, insects, etc. blocking air flow.

4. Check radiator and hoses for leaks, loose connections, loose mountings, corrosion, chafing, and missing parts.

5. Check coolant level at coolant recovery (overflow) bottle.

6. Close both engine access doors.

END OF TASK

SERVICING

WARNING

Cooling system operates at high temperature and pressure. Contact with high pressure steam and/ or liquids can result in burns and scalding. Shut down generator set, and allow system to cool before performing checks, services and maintenance. Failure to comply can cause injury or death to personnel.

Table 1. Coolant.

COOLANT			
AMBIENT TEMPERATURE	**RADIATOR COOLANT**		**RATIO**
+40°FTO+120°F (+4°CTO+49°C)	Water:	MIL-A-53009A (1) INHIBITOR, CORROSION	35:1
-25°FTO+120°F (-32°CTO+49°C)	Water:	A-A-52624A ANTIFREEZE	1:1
-25°FTO+120°F (-32°CTO+49°C)	Water:	A-A-52624A ANTIFREEZE	NA

1. Shut down generator set. Refer to WP 0005, Stopping Procedure.

2. Open left side engine access door.

3. Remove cap on coolant recovery (overflow) bottle.

4. Fill coolant recovery (overflow) bottle to HOT line if coolant is hot or to COLD line if coolant is cold with proper coolant/antifreeze in accordance with Table 1.

5. Install coolant recovery (overflow) bottle cap.

6. Close left side engine access door.

END OF TASK

END OF WORK PACKAGE

OPERATOR MAINTENANCE

15 kW 50/60 AND 400 Hz SKID MOUNTED, TACTICAL QUIET GENERATOR SET

FUEL TANK: INSPECTION, SERVICE

INITIAL SETUP:

Tools and Special Tools
Generator Mechanical Tool Kit

Materials/Parts
Diesel/Turbine Fuel A-A-52557A Grade 2-D
MIL-DTL-831-33E, JP-8
Diesel/Turbine Fuel A-A-52557A Grade 1-D
MIL-DTL-5624T, JP-5

Personnel Required
One

References
WP 0005, Stopping Procedures

Equipment Condition
Grounded off & operational

INSPECTION

WARNING

Fuels used in the generator set are combustible. Do not smoke or use open flames when performing maintenance. Failure to comply can result in flames and possible explosion and can cause injury or death to personnel and damage to the generator set.

1. Place MASTER SWITCH in PRIME & RUN or PRIME & RUN AUX FUEL position.
2. Check fuel level by observing FUEL LEVEL indicator.
3. Remove fuel cap and ensure strainer is free of dirt and other foreign material.

END OF TASK

SERVICING

CAUTION

Use only specified diesel fuel to service the fuel tank. Refer to Table 1. Otherwise, equipment damage could result.

1. Shut down generator set. Refer to WP 0005, Stopping Procedure.
2. Remove fuel cap.
3. Remove fuel strainer, clean as necessary, and reinstall.

NOTE

Fuel tank holds 14 gallons (53 liters).

4. Add diesel fuel to fuel tank.
5. Install fuel cap.

Table 1. Diesel Fuel.

FUEL	
AMBIENT TEMPERATURE	DIESEL/TURBINE FUEL

Table 1. Diesel Fuel. - Continued

FUEL	
AMBIENT TEMPERATURE	DIESEL/TURBINE FUEL
+20°FTO+120°F (-7°CTO+49°C)	A-A-52557A, GRADE 2-D MIL-DTL-83133E, JP-8
-25°FTO+20°F (-32°CTO+7°C)	A-A-52557A, GRADE 1-D MIL-DTL-5624T, JP-5

END OF TASK

END OF WORK PACKAGE

OPERATOR MAINTENANCE

15 kW 50/60 AND 400 Hz SKID MOUNTED, TACTICAL QUIET GENERATOR SET
FUEL FILTER/WATER SEPARATOR: INSPECTION, SERVICE

INITIAL SETUP:

Tools and Special Tools Generator Mechanical Tool Kit	**Personnel Required** One
References WP 0005, Stopping Procedure	**Equipment Condition** Grounded off & operational

INSPECTION

1. Shut down generator set. Refer to WP 0005, Stopping Procedure.

2. Open right side engine access door.

3. Inspect fuel filter/water separator assembly for proper mounting cracks, dents, leaks, loose fuel lines and other damage.

4. Close right side engine access door.

END OF TASK

SERVICING

1. Shut down generator set. Refer to WP 0005, Stopping Procedure.

2. Open right side engine access door.

3. Open fuel drain cock (Figure 1, Item 2) and air vent (1) on fuel filter/water separator assembly and drain any sediment and water into a suitable container.

4. Close drain cock (2) and air vent (1).

5. Close right side engine access door.

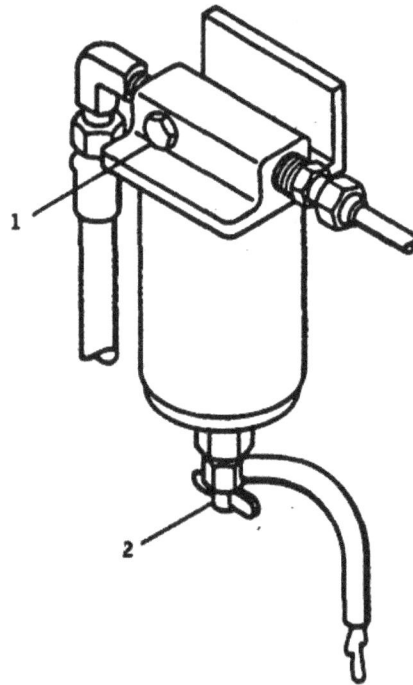

Figure 1. Draining Fuel Filter/Water Separator.

END OF TASK

END OF WORK PACKAGE

OPERATOR MAINTENANCE

15 kW 50/60 AND 400 Hz SKID MOUNTED, TACTICAL QUIET GENERATOR SET

LUBRICATION SYSTEM: INSPECTION, SERVICE

INITIAL SETUP:

Tools and Special Tools
Generator Mechanical Tool Kit

Materials/Parts
Oil

Personnel Required
One

References
WP 0005, Stopping Procedure
LO 9-6115-643-12

Equipment Condition
Generator set grounded, off & operational

INSPECTION

1. Shut down generator set. Refer to WP 0005, Stopping Procedure.

2. Open both engine access doors.

3. Inspect engine assembly for oil leaks.

4. Check for damage, proper mounting, or missing parts.

CAUTION

The dipstick is marked so that the crankcase oil can be checked while engine is stopped or running. Always make sure correct side of dipstick is checked. Remove oil filler cap when checking oil with engine running.

5. Check engine crankcase oil level. Refer to LO 9-6115-643-12.

6. Close both engine access doors.

END OF TASK

SERVICING

1. Shut down generator set. Refer to WP 0005, Stopping Procedure.

2. Open left side engine access door (MEP-804A/MEP-814A) or right side engine access door (MEP-804B/MEP-814B).

3. Remove oil filter cap.

4. Add oil to engine crankcase. Refer to LO 9-6115-643-12.

5. Install oil filter cap.

6. Close side engine access door.

END OF TASK

END OF WORK PACKAGE

OPERATOR MAINTENANCE

15 kW 50/60 AND 400 Hz SKID MOUNTED, TACTICAL QUIET GENERATOR SET
CRANKCASE VENTILATION FILTER (MEP-804B/MEP-814B): INSPECTION, SERVICE

INITIAL SETUP:

Tools and Special Tools
    Generator Mechanical Tool Kit

Personnel Required
    One

Equipment Condition
    Generator set grounded, off & operational

Materials/Parts
    Warm, soapy water
    Filter element

References
    WP 0005, Stopping Procedure

INSPECTION

1. Shut down generator set. Refer to WP 0005, Stopping Procedure.

2. Open right side engine access door.

3. Inspect crankcase ventilation filter assembly for proper mounting, cracks, dents, leaks, loose oil lines/hoses and other damage.

4. Close right side engine access door.

END OF TASK

SERVICING

1. Shut down generator set. Refer to WP 0005, Stopping Procedure.

2. Open right side engine access door.

3. Release two spring latches (Figure 1, Item 1) and remove cover (2).

4. Remove filter element (3).

5. Clean filter element (3) with warm soapy water and rinse and dry thoroughly.

6. Install filter element (3) into cover (2).

7. Install cover (2) and secure with two spring latches (1).

8. Close right side engine access door.

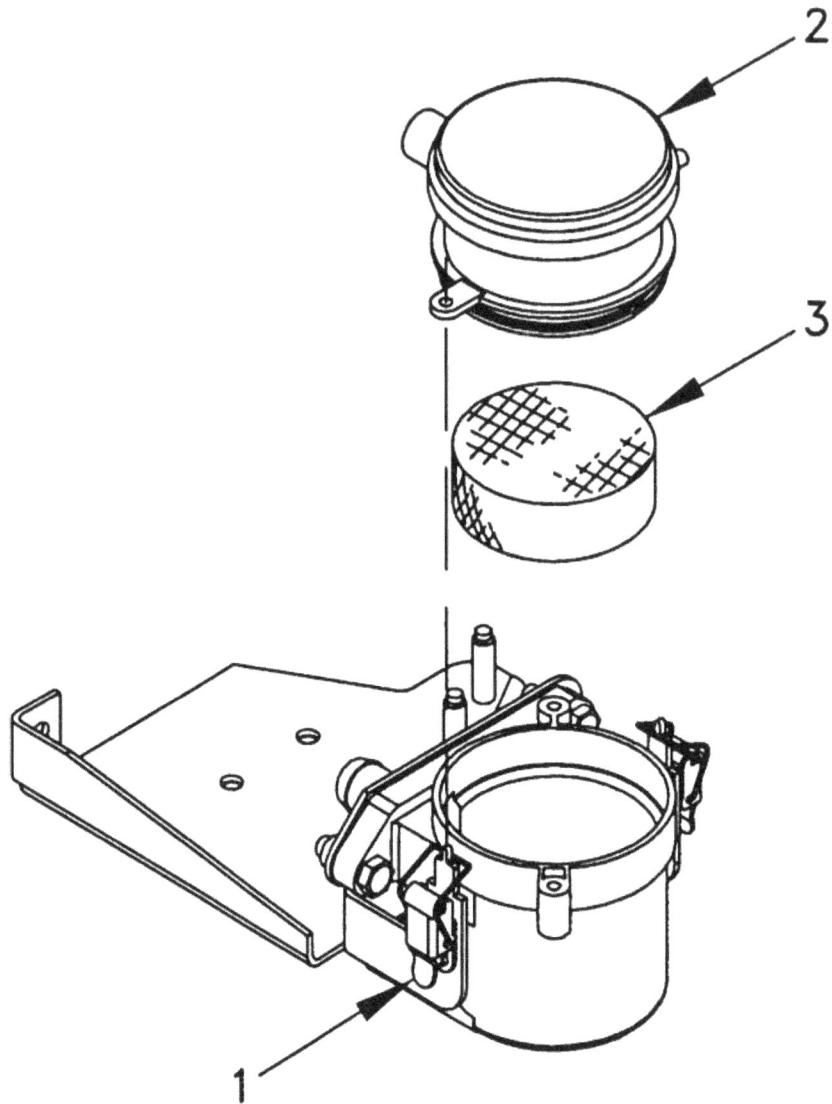

Figure 1. Crankcase Ventilation Filter - MEP-804B/MEP-814B.

END OF TASK

END OF WORK PACKAGE

CHAPTER 5

OPERATOR SUPPORTING INFORMATION

FOR

15 kW 50/60 AND 400 Hz SKID MOUNTED, TACTICAL QUIET GENERATOR SET

CHAPTER 5

SUPPORTING INFORMATION

WORK PACKAGE INDEX

OPERATOR MAINTENANCE

15 kW 50/60 AND 400 Hz SKID MOUNTED, TACTICAL QUIET GENERATOR SET

REFERENCES

SCOPE

This work package lists all forms, regulations, pamphlets, specifications, standards, technical manuals, technical bulletins, lubrication orders, field manuals, and miscellaneous publications referenced in this TM.

FORMS

DA Form 2028 Recommended Changes to Publications and Blank Forms

DA Form 2028-2 Recommended Changes to Equipment Technical Publications

DA Form 2404 Equipment Inspection and Maintenance Worksheet

DA Form 2407 Maintenance Request

DA Form 2408 Equipment Log Assembly (Records)

DA Form 2408-9 Equipment Control Record

DA Form 2408-20 Oil Analysis Log

DA Form 5988-E Equipment Inspection and Maintenance Worksheet

DD Form 314 Preventive Maintenance Schedule and Record

SF Form 364 Report of Discrepancy

SF Form 368 Product Quality Deficiency Report

ARMY REGULATIONS

AR 310-25 Dictionary of United States Army Terms

DEPARTMENT OF THE ARMY PAMPHLETS

DA PAM 750-8 The Army Maintenance Management System (TAMMS)

MILITARY SPECIFICATIONS

MIL-DTL-5624T Turbine Fuel, Aviation, Grades JP-4, JP-5, and JP-5/JP-8 ST

MIL-A-53009A(1) Additive, Antifreeze Extender, Liquid Cooling Systems

MIL-DTL-83133E Turbine Fuels, Aviation, Kerosene Types, NATO F-34 (JP-8), NATO F-35 and JP-8+100

COMMERCIAL ITEM DESCRIPTIONS

A-A-52557A Fuel Oil, Diesel; for Posts, Camps, and Stations

A-A-52624A Antifreeze, Multi Engine Type

ASME-Y14.38M Abbreviations for Use on Drawings, and in Specifications, Standards and Technical Documents

MILITARY STANDARDS

None   N/A

TECHNICAL MANUALS

TM 750-244-3 Procedures for Destruction of Equipment to Prevent Enemy Use (Mobility Equipment Command)

TECHNICAL BULLETINS

TB 43-0125 Installation of Communications Electronic Equipment: Hookup of Electrical Cables to Mobile Generator Sets on Fielded Equipment to Meet Electrical Safety Standards

LUBE ORDERS

LO 9-6115-643-12 Generator Set, Skid Mounted, Tactical Quiet 15 kW, 50/60 and 400 Hz MEP-804A, Tactical Quiet, 50/60 Hz, NSN 6115-01-274-7388 MEP-804B, Tactical Quiet, 50/60 Hz, NSN 6115-01-530-1458 MEP-814A, Tactical Quiet, 400 Hz, NSN 6115-01-274-7393 MEP-814B, Tactical Quiet, 400 Hz, NSN 6115-01-529-9494

FIELD MANUALS

FM 3-3 Chemical and Biological Contamination Avoidance

FM 3-4 NBC Protection

FM 3-5 NBC Decontamination

FM 4-25.11 First Aid

FM 5-424 Theater of Operations, Electrical Systems

FM 9-207 Operation and Maintenance of Ordnance Materiel in Cold Weather (0 ° to −65 °)

FM 21-6 Techniques of Military Instruction

FM 21-30 Military Symbols

FM 21-40 Chemical, Biological, Radiological, and Nuclear Defense

FM 31-70 Basic Cold Weather Manual

FM 31-71 Northern Operations

FM 90-6 Mountain Operations

MISCELLANEOUS PUBLICATIONS

AFR 66-1 Air Force Maintenance Forms and Records

AR 700-138 Army Logistics Readiness and Sustainability

AR 735-11-2 Reporting of Supply Discrepancies

AR 750-1 Army Materiel Maintenance Policy and Retail Maintenance Operations

AR 750-244-2 Procedures for Destruction of Electronics Materiel to Prevent Enemy Use

CTA 8-100 Army Medical Department Expendable/Durable Items

CTA 50-970 Expendable Items (Except Medical Class V, Repair Parts, and Heraldic Items)

END OF WORK PACKAGE

OPERATOR MAINTENANCE

15 kW 50/60 AND 400 Hz SKID MOUNTED, TACTICAL QUIET GENERATOR SET
COMPONENTS OF END ITEM (COEI) AND BASIC ISSUE ITEMS (BII) LISTS

COMPONENTS OF END ITEM (COEI) AND BASIC ISSUE ITEMS (BII) LISTS

INTRODUCTION

Scope

This work package lists COEI and BII for the Generator Set to help you inventory items for safe and efficient operation of the equipment.

General

The COEI and BII information is divided into the following lists:

Components of End Item (COEI). This list is for information purposes only and is not authority to requisition replacements. These items are part of the Generator Set. As part of the end item, these items must be with the end item whenever it is issued or transferred between property accounts. Items of COEI are removed and separately packaged for transportation or shipment only when necessary. Illustrations are furnished to help you find and identify the items.

Basic Issue Items (BII). These essential items are required to place the Generator Set in operation, operate it, and to do emergency repairs. Although shipped separately packaged, BII must be with the Generator Set during operation and when it is transferred between property accounts. Listing these items is your authority to request/requisition them for replacement based on authorization of the end item by the TOE/MTOE. Illustrations are furnished to help you find and identify the items.

Explanation of Columns in the COEI List and BII List

Column (1) Illus Number. Gives you the number of the item illustrated.

Column (2) National Stock Number (NSN). Identifies the stock number of the item to be used for requisitioning purposes.

Column (3) Description, Part Number/(CAGEC). Identifies the Federal item name (in all capital letters) followed by a minimum description when needed. The stowage location of COEI and BII is also included in this column. The last line below the description is the part number and the Commercial and Government Entity Code (CAGEC) (in parentheses).

Column (4) Usable On Code. When applicable, gives you a code if the item you need is not the same for different models of equipment.

Column (5) U/I. Unit of Issue (U/I) indicates the physical measurement or count of the item as issued per the National Stock Number shown in column (2).

Column (6) Qty Rqr. Indicates the quantity required.

There is no COEI table for this TM.

LUBRICATION ORDER	LO 9-6115-643-12
31 JULY 2008	This Lubrication Order supersedes LO 9-6115-643-12, dated 30 October 1996

GENERATOR SET, SKID MOUNTED
TACTICAL QUIET 15kW, 50/60 AND 400 Hz

DOD MODEL	CLASS	HERTZ	NSN
MEP-804A	TACTICAL QUIET	50/60	6115-01-274-7388
MEP-814A	TACTICAL QUIET	400	6115-01-274-7393
MEP-804B	TACTICAL QUIET	50/60	6115-01-530-1458
MEP-814B	TACTICAL QUIET	400	6115-01-529-9494

Reference TM 9-6115-643-10

REPORTING ERRORS AND RECOMMENDING IMPROVEMENTS

You can help improve this LO. If you find any mistakes or if you know of a way to improve the procedures, please let us know. Reports, as applicable by the requiring Service, should be submitted as follows:

Mail your letter or DA Form 2028 (Recommended Changes to Publications and Blank Forms) located in the back of this manual, directly to: Commander, U.S. Army CECOM Life Cycle Management Command (LCMC) and Fort Monmouth, ATTN: AMSEL-LC-LEO-E-ED, Fort Monmouth, NJ 07703-5006. You may also send in your recommended changes via electronic mail or by fax. Our fax number is 732-532-1556, DSN 992-1556. Our e-mail address is MONM-AMSELLEOPUBSCHG@conus.army.mil. Our online web address for entering and submitting DA Form 2028s is http://edm.monmouth.army.mil/pubs/2028.html.

A reply will be furnished to you.

"Copy of this Lubrication Order will remain with the equipment at all times. Instructions contained herein are mandatory."

1

Figure 1. Item 1. Basic Issue Items Lubrication Order LO 9-6115-643-12.

*ARMY TM 9-6115-643-10
AIR FORCE TO 35C2-3-445-21

TECHNICAL MANUAL

OPERATOR'S MANUAL

FOR

**GENERATOR SET, SKID MOUNTED, TACTICAL QUIET,
15 kW, 50/60 Hz, MEP-804A
(NSN: 6115-01-274-7388) (EIC: VG4)
15 kW, 50/60 Hz, MEP-804B
(NSN: 6115-01-530-1458) (EIC: N/A)**

**GENERATOR SET, SKID MOUNTED, TACTICAL QUIET,
15 kW, 400 Hz, MEP-814A
(NSN: 6115-01-274-7393) (EIC: VN4)
15 kW, 400 Hz, MEP-814B
(NSN: 6115-01-529-9494) (EIC: N/A)**

*SUPERSEDURE NOTICE - This manual supersedes TM 9-6115-643-10 dated 01 April 2008. Date of issue for the revised manual is: 15 February 2010.

**HEADQUARTERS, DEPARTMENTS OF THE ARMY
AND AIR FORCE**
15 FEBRUARY 2010

Figure 2. Item 2. Basic Issue Items Technical Manual TM 9-6115-643-10.

* TB 9-6115-643-24

DEPARTMENT OF THE ARMY TECHNICAL BULLETIN

WARRANTY PROGRAM
FOR
GENERATOR SET, TACTICAL QUIET
15 KW, 50/60 AND 400 HZ
MEP-804A, MEP-804B, MEP-814A AND MEP-814B

Headquarters, Department of the Army, Washington, D.C.

1 August 2008

* This bulletin supersedes TB 9-6115-643-24, dated 30 October 1996

REPORTING ERRORS AND RECOMMENDING IMPROVEMENTS

You can help improve this manual. If you find any mistakes or if you know of a way to improve the procedures, please let us know. Reports, as applicable by the requiring Service, should be submitted as follows.

Mail your letter or DA Form 2028 (Recommended Changes to Publications and Blank Forms) located in the back of this manual, directly to: Commander, U.S. Army CECOM Life Cycle Management Command (LCMC) and Fort Monmouth, ATTN: AMSEL-LC-LEO-E-ED, Fort Monmouth, NJ 07703-5006. You may also send in your recommended changes via electronic mail or by fax. Our fax number is 732-532-1556, DSN 992-1556. Our e-mail address is MONM-AMSELLEOPUBSCHG@conus.army.mil. Our online web address for entering and submitting DA Form 2028s is http://edm.monmouth.army.mil/pubs/2028.html.

A reply will be furnished to you.

DISTRIBUTION STATEMENT A: Approved for public release; distribution is unlimited

Figure 3. Item 3. Basic Issue Items Warranty Technical Bulletin TB 9-6115-643-24.

Table 1. Basic Issue Items List.

(1) Illus Number	(2) National Stock Number (NSN)	(3) Description, Part Number / (CAGEC)	(4) Usable On Code	(5) U/I	(6) Qty Rqr
1		LUBRICATION ORDER LO 9-6115-643-12		EA	1
2		TECHNICAL MANUAL TM 9-6115-643-10		EA	1
3		WARRANTY TECHNICAL BULLETIN TB 9-6115-643-24		EA	1

END OF WORK PACKAGE

OPERATOR MAINTENANCE

15 kW 50/60 AND 400 Hz SKID MOUNTED, TACTICAL QUIET GENERATOR SET

ADDITIONAL AUTHORIZATION LIST (AAL)

ADDITIONAL AUTHORIZATION LIST (AAL)

INTRODUCTION

Scope

This work package lists additional items you are authorized for the support of the Generator Set.

General

This list identifies items that do not have to accompany the Generator Set and that do not have to be turned in with it. These items are all authorized to you by CTA, MTOE, TDA, or JTA.

Explanation of Entries in the AAL

Column (1) National Stock Number (NSN). Identifies the stock number of the item to be used for requisitioning purposes.

Column (2) Description, Part Number/(CAGEC). Identifies the Federal item name (in all capital letters) followed by a minimum description when needed. The last line below the description is the part number and the Commercial and Government Entity Code (CAGEC) (in parentheses).

Column (3) Usable On Code. When applicable, gives you a code if the item you need is not the same for different models of equipment.

Column (4) U/I. Unit of Issue (U/I) indicates the physical measurement or count of the item as issued per the National Stock Number shown in column (1).

Column (5) Qty Recm. Indicates the quantity recommended.

Table 1. Additional Authorization List.

(1) National Stock Number (NSN)	(2) Description, Part Number / (CAGEC)	(3) Usable On Code	(4) U/I	(5) Qty Recm
5342-00-066-1235	ADAPTER, CONTAINER 13211E7541 (97403)		EA	1
4210-00-361-6921	EXTINGUISHER, FIRE, CARBON DIOXIDE, 5 LB 322 (54905)		EA	1
7240-00-177-6154	FLEXIBLE SPOUT MIL-S-1285 (81349)		EA	1
7240-01-337-5269	FUEL CAN		EA	1
5120-01-013-1676	HAMMER, SLIDE, GROUND 0116-1810 (93742)		EA	1

END OF WORK PACKAGE

OPERATOR MAINTENANCE

15 kW 50/60 AND 400 Hz SKID MOUNTED, TACTICAL QUIET GENERATOR SET
EXPENDABLE AND DURABLE ITEMS LIST

EXPENDABLE AND DURABLE ITEMS LIST

Scope

This work package lists expendable and durable items that you will need to operate and maintain the (enter equipment/end item name). This list is for information only and is not authority to requisition the listed items. These items are authorized to you by CTA 50-970, Expendable/Durable Items (Except Medical, Class V Repair Parts, and Heraldic Items), CTA 50-909, Field and Garrison Furnishings and Equipment or CTA 8-100, Army Medical Department Expendable/Durable Items.

Explanation of Columns in the Expendable/Durable Items List

Column (1) Item No. This number is assigned to the entry in the list and is referenced in the narrative instructions to identify the item (e.g., Use brake fluid (WP 0098, Item 5)).

Column (2) Level. This column identifies the lowest level of maintenance that requires the listed item (include as applicable: C = Crew, O = AMC, F = Maintainer or ASB, H = BelowDepot or TASMG, D = Depot).

Column (3) National Stock Number (NSN). This is the NSN assigned to the item which you can use to requisition it.

Column (4) Item Name, Description, Part Number/(CAGEC). This column provides the other information you need to identify the item. The last line below the description is the part number and the Commercial and Government Entity Code (CAGEC) (in parentheses).

Column (5) U/I. Unit of Issue (U/I) code shows the physical measurement or count of an item, such as gallon, dozen, gross, etc.

Table 1. Expendable and Durable Items List.

(1) Item No.	(2) Level	(3) National Stock Number (NSN)	(4) Item Name, Description, Part Number / (CAGEC)	(5) U/I
1	F	8040-00-664-4318	Adhesive 9995460 (18876)	EA
2	F	6850-00-181-7929	Antifreeze A-A-52624 (81349)	GL
3	F	6850-01-331-3349	Cleaning compound, solvent P-D-680 (81348)	EA
4	F	6850-01-331-3350	Cleaning compound, solvent P-D-680 (81348)	EA
5	F	7920-01-338-3329	Cloth, Cleaning	EA
6	F	9150-00-190-0904	Grease, Automotive/artillery GAA MIL-PRF-10924 (81349)	EA
7	F	9150-00-189-6727	Oil, Lubrication OE/HDO-10 MIL-PRF-2104 (81349)	EA
8	F		Solder Sn60Pb40 (81348)	EA
9	F	6810-00-107-1510	Water, Distilled	GL

END OF WORK PACKAGE

CHAPTER 6

OPERATOR SUPPORTING INFORMATION

FOR

WINTERIZATION KIT

CHAPTER 6

WINTERIZATION KIT

WORK PACKAGE INDEX

OPERATOR MAINTENANCE

WINTERIZATION KIT

WINTERIZATION KIT, GENERAL INFORMATION

SCOPE

The winterization kit is designed to be mounted in generator sets where extreme cold temperatures are anticipated. The kit contains a coolant heater that allows the generator set to operate to -50°F (-45.6°C). The kit heater pump circulates the generator set coolant through the heater pump, heats the coolant and then returns the coolant back through the radiator of the generator set. This cycle continues in high heat mode until the temperature reaches 176°F(80°C). The heater then switches into a low heat mode. If the coolant temperature drops to 158°F(70°C) the heater will automatically switch to the high heat mode.

END OF WORK PACKAGE

OPERATOR MAINTENANCE

WINTERIZATION KIT

WINTERIZATION KIT, EQUIPMENT DESCRIPTION AND DATA

EQUIPMENT CHARACTERISTICS, CAPABILITIES AND FEATURES

Characteristics

The Winterization Kit contains a coolant heater that heats the coolant and allows the generator set to operate to -50°F (-45.6°C).

Capabilities and Features

The heater burns fuel from the generator set fuel tank to heat the coolant that is pumped back through the engine block. The kit consists of a heater and coolant pump, a control unit, an ON-OFF switch, a fuel pump and line, coolant circulating lines, bypass valve, a wiring harness and mounting hardware to ensure operation to -50°F (-45.6°C).

LOCATION AND DESCRIPTION OF MAJOR COMPONENTS

Figure 1 illustrates the major components of the kit and shows their locations on the 15 kW Tactical Quiet Generator (TQG) Set. (Refer to Table 1 for item names).

Table 1. Description of Major Winterization Kit Components.

Item No.	Item Name	Description
	Winterization Kit	A fuel-burning heater, pre-heats engine coolant permitting generator set Op-er-ation to -50°F (-45.6°C).
1	Control Unit	Controls heater operations.
2	Heater	Heats coolant for operation in extreme cold temperatures.
3	Fuel Pump	Pumps fuel from the generator set fuel tank to the heater.
4	Fuel Lines	Provides a means of transporting fuel to heater.
5	Coolant Pump	Circulates coolant from generator set through the heater.
6	Coolant Lines	Provides a means of transporting coolant for circulation.
7	Bypass Valve	Allows coolant to bypass heater when Winterization Kit is not in use. (MEP-804B/MEP-814B Only)
8	Switch/Lamp	Switches heater on or off and lamp indicates heater function codes.
9	Wiring Harness	Electrically connects Winterization Kit components.
10	Exhaust Hose	Provides a means of exhausting combustion gases from heater.
11	Air Inlet Hose	Provides intake air to winterization heater.

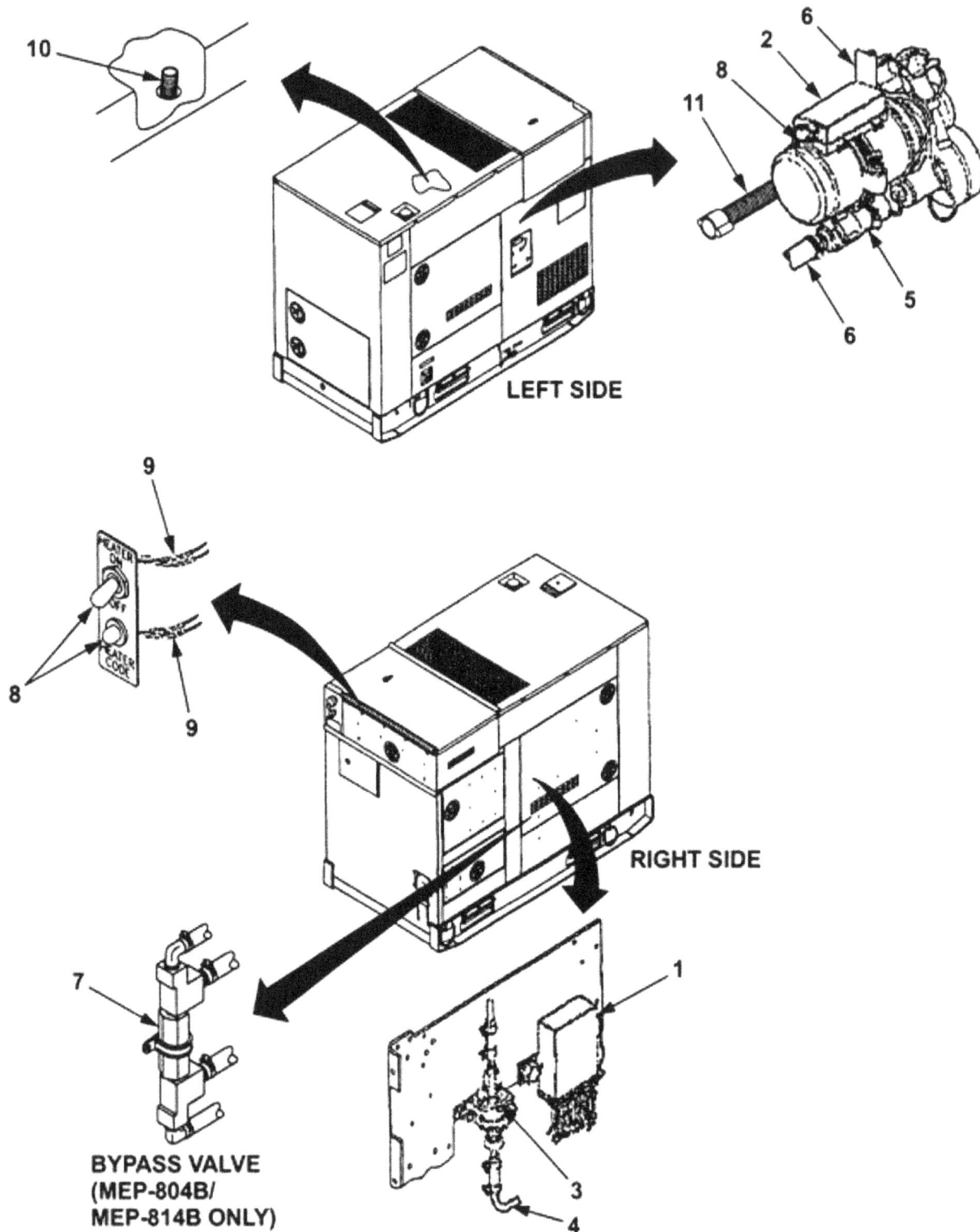

Figure 1. Location of Major Winterization Kit Components.

TABULATED/ILLUSTRATED DATA

Tabulated data for the heater is located in Table 2.

Table 2. Heater Operating Data.

Item Name	Data
1. Winterization Kit	
a. National Stock Numberb. Overall Lengthc. Overall Widthd. Overall Heighte. Weight	6115-01-477-0566
	10.787 in
	5.984 in
2. Heater	7.815 in
a. Manufacturer	15 lbs.
b. Model	
3. Heating	
Capacity	Active Gear
	D5W
	Water Coolant
4. Rated Voltage	High: 17,000 BTU/Hr.
a. Operating Voltage Rangeb. Current at 24 VDC	Low: 4250 BTU/Hr.
	24 VDC
	20-28 VDC
5. Fuel	Start: 20 Amps/Hr.
Fuel Consumption	Running High: 1.8 Amps/Hr.
	Running Low: 1.2 Amps/Hr.
6. Coolant Pump Flow	Diesel
	High: 0.16 Gal/Hr.
	Low: 0.04 Gal/Hr.
	250 Gal/Hr.

END OF WORK PACKAGE

OPERATOR MAINTENANCE

15 kW 50/60 AND 400 Hz SKID MOUNTED, TACTICAL QUIET GENERATOR SET

WINTERIZATION KIT, TROUBLESHOOTING PROCEDURES

GENERAL

Refer to Chapter 3, WP 0009 for generator set troubleshooting procedures. This work package lists common malfunctions you may find during operation of the generator set with the Winterization Kit installed and the generator set is running. You should perform the tests/inspections and corrective actions in the order listed. The troubleshooting symptom index cannot list all faults that may occur, nor all the tests or inspections and corrective actions. If a malfunction is not listed or cannot be corrected by listed corrective actions, notify your supervisor.

Code Light Troubleshooting

The indicator light near the heater switch is designed to blink on codes sequences to signal malfunctions in the system. (Refer to Code Light Pulses)

Code Light Pulses

The indicator light near the heater ON-OFF switch will blink in different sequences of long and short to indicate malfunctions. A plate (Figure 1) mounted on the generator control panel access door lists the malfunctions and shows each sequence of pulses. If you see any of these series of pulses, notify the next-higher level of maintenance.

NOTE

Before performing troubleshooting procedures, turn off heater and attempt restart.

SYMPTOM INDEX, WINTERIZATION KIT

NOTE

When the heater is switched on, the light will perform one of the sequences of light pulses shown visually on the Function Codes Plate mounted inside the generator control panel cover (Figure 1). Before each symptom, this index lists in parentheses the light sequence associated with it.

(long dash, short dash, long dash) - Start, glow period

(continuous dash) - Normal Function

(long dash, long dash) - Purge Cycle and Restart

(dash, dash) - Heater Restart attempted During Purge Cycle

(dash, 5 dots, dash) - Warning: Power supply

(10 dots) -Overheating

(dot, dot) - Flame Sensor Short-circuit

(2 dots, 2 dots) - Flame Cutout-LOW

(3 dots, 3 dots) - Flame Cutout-HIGH

(4 dots, 4 dots) - Glow Plug Defect

(dash, dash) - Burner Motor Defect

(dash, dot, dash, dot) - Under voltage

(dash, 2 dots, dash, 2 dots) - Over voltage

(dash, 3 dots, dash, 3 dots) - Non-start

(2 dots, dash, 2 dots, dash) - Temperature Sensor Defective

(3 dots, dash, 3 dots, dash) - Fuel pump short circuit

(2 dots, dash, 3 dots, dash, dot) - Temperature switch defective

(4 dashes) - Control unit defective

(dot, dash, 3 dots, dash, 2 dots) - Connection error

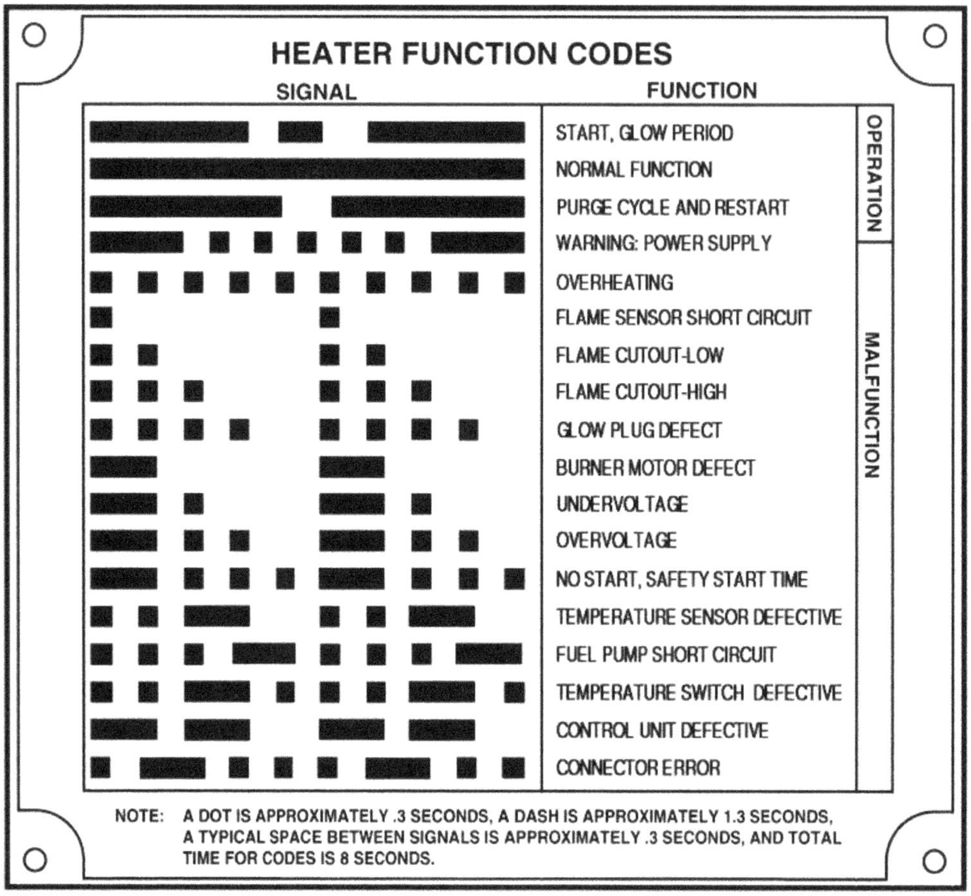

Figure 1. Heater Function Codes Plate.

END OF WORK PACKAGE

OPERATOR MAINTENANCE

15 kW 50/60 AND 400 Hz SKID MOUNTED, TACTICAL QUIET GENERATOR SET

WINTERIZATION KIT, PREVENTIVE MAINTENANCE CHECKS AND SERVICES (PMCS) INTRODUCTION

INTRODUCTION TO OPERATOR PMCS TABLE

WP 0027 (PMCS Table 1) has been provided so you can keep your equipment in good operating condition and ready for its primary mission.

Warnings, Cautions, and Notes

Always observe the**WARNINGS, CAUTIONS,**and**NOTES**appearing in your PMCS table. Warnings and cautions appear before applicable procedures. You must observe **WARNINGS**to prevent serious injury to yourself and others. You must observe **CAUTIONS**to prevent your equipment from being damaged. You must observe**NOTES**to ensure procedures are performed properly.

Explanation of Table Entries

The PMCS table is divided into five columns. Each column is explained in the following paragraphs.

Item No. Column. Numbers in this column are for reference. When completing DA Form 2404 (Equipment Inspection and Maintenance Worksheet), include the item number for the check/service indicating a fault. Item numbers also appear in the order that you must do Checks and Services for the intervals listed.

Interval Column. This column tells you when you must do the procedure described in the procedure column. "BEFORE" procedures must be done before you operate the equipment for its intended mission. "DURING" procedures must be done during the time you are operating the equipment for its intended mission. "AFTER" procedures must be done immediately after you have operated the equipment. Perform "WEEKLY" procedures at the listed interval.

Location, Item to Check/Service Column. This column lists the location and the item to be checked or serviced. The item location is underlined.

Procedure Column. This column gives the procedure for checking or servicing the item listed in the location, item to check/service column. You must perform the procedure to know if the power unit or power plant is ready or available for its intended mission or operation. You must do the procedure at the time stated in the interval column.

Equipment Not Ready/Available if: Column. Information in this column tells you what faults will keep your equipment from being capable of performing its primary mission. If you perform checks or services that show faults listed in this column, do not operate the equipment.

Reporting and Correcting Deficiencies

If Winterization Kit does not perform as required, refer to Chapter 3, Operator Troubleshooting Procedures.

Other Table Entries

Be sure to observe all special information and notes that appear in your table.

Special Instructions

Preventive maintenance is not limited to performing the Checks and Services listed in the PMCS table. Covering unused receptacles, stowing unused accessories, and performing other routine procedures such as equipment inventory, cleaning components, and touch-up painting are not listed in the table. These are things you should do any time you see that they need to be done. If a routine check is listed in the PMCS table, it is because experience has shown that problems may occur with this item. Take along tools and cleaning cloths needed to perform the required Checks and Services. Use the information in the following paragraphs to help you identify problems at any time and to help identify potential problems before and during Checks and Services.

WARNING

Metal jewelry can conduct electricity and become entangled in generator set components. Remove all metal jewelry when working on generator set. Failure to comply can cause injury or death to personnel.

WARNING

Do not wear loose clothing when performing checks, services and maintenance. Loose clothing may be entangled in generator set components. Failure to comply can cause injury or death to personnel.

WARNING

High voltage is produced when this generator set is in operation. Make sure unit is completely shut down and free of any power source before attempting any repair or maintenance on the unit. Failure to comply can cause injury or death to personnel.

WARNING

Solvent used to clean parts is potentially dangerous to personnel and property. Clean parts in a well-ventilated area. Avoid inhalation of solvent fumes. Wear goggles and rubber gloves to protect eyes and skin. Wash exposed skin thoroughly. Do not smoke or use near open flame or excessive heat. Failure to comply with this warning can cause injury to personnel, and damage to the equipment.

CAUTION

Keep cleaning solvents, fuels and lubricants away from rubber or soft plastic parts. They will deteriorate material.

1. Keep the generator set clean. Dirt, grease, and oil get in the way and may cover up a serious problem. Use cleaning solvent to clean metal surfaces.
2. Use soap and water to clean rubber or plastic parts and material.
3. Check all bolts, nuts, and screws to make sure they are not loose, missing, bent, or broken. Do not try to check them with a tool, but look for chipped paint, bare metal, or rust around bolt heads. If you find one loose, report it to the next-higher level of maintenance.
4. Inspect welds for loose or chipped paint, rust, or gaps where parts are welded together. If a broken weld is found, report it to the next-higher level of maintenance.
5. Inspect electrical wires, connectors, terminals, and receptacles for cracked or broken insulation, bare wires, and loose or broken connectors. Tighten loose connectors. Examine terminals and receptacles for serviceability. If deficiencies are found, report them to the next-higher level of maintenance.
6. Inspect hoses and fluid lines. Look for wear, damage, and leaks. Make sure that clamps and fittings are tight. Wet spots and stains around a fitting or connector can mean a leak. If a leak comes from a loose connector, or if something is broken or worn out, report it to the next-higher level of maintenance.

Leakage Definitions

You must know how fluid leakage affects the status of your equipment. The following are definitions of the types/ classes of leakage you need to know to be able to determine the status of your equipment. Learn and be familiar with them. When in doubt, notify your supervisor.

Table 1. Leakage Definitions.

Leakage Class Leakage Definition

Class I Seepage of fluid (as indicated by wetness or discoloration) not great enough to form drops.

Class II Leakage of fluid great enough to form drops, but not enough to cause drops to drip from the item being checked/inspected.

Class III Leakage of fluid (other than fuel) greater than three drops per minute that fall from the item being inspected.

Order in Which PMCS Will be Done

Figure 1 shows the order in which you are to perform your PMCS. The figure shows a generator set to which a kit has been added. The number call outs on Figure 1 correspond to the numbers in the Item No. column of WP 0027, Table 1 (for BEFORE/DURING PMCS).

NOTE

Be sure Generator Set PMCS is completed first in accordance with Chapter 4, Maintenance Instructions, WP 0011, PMCS.

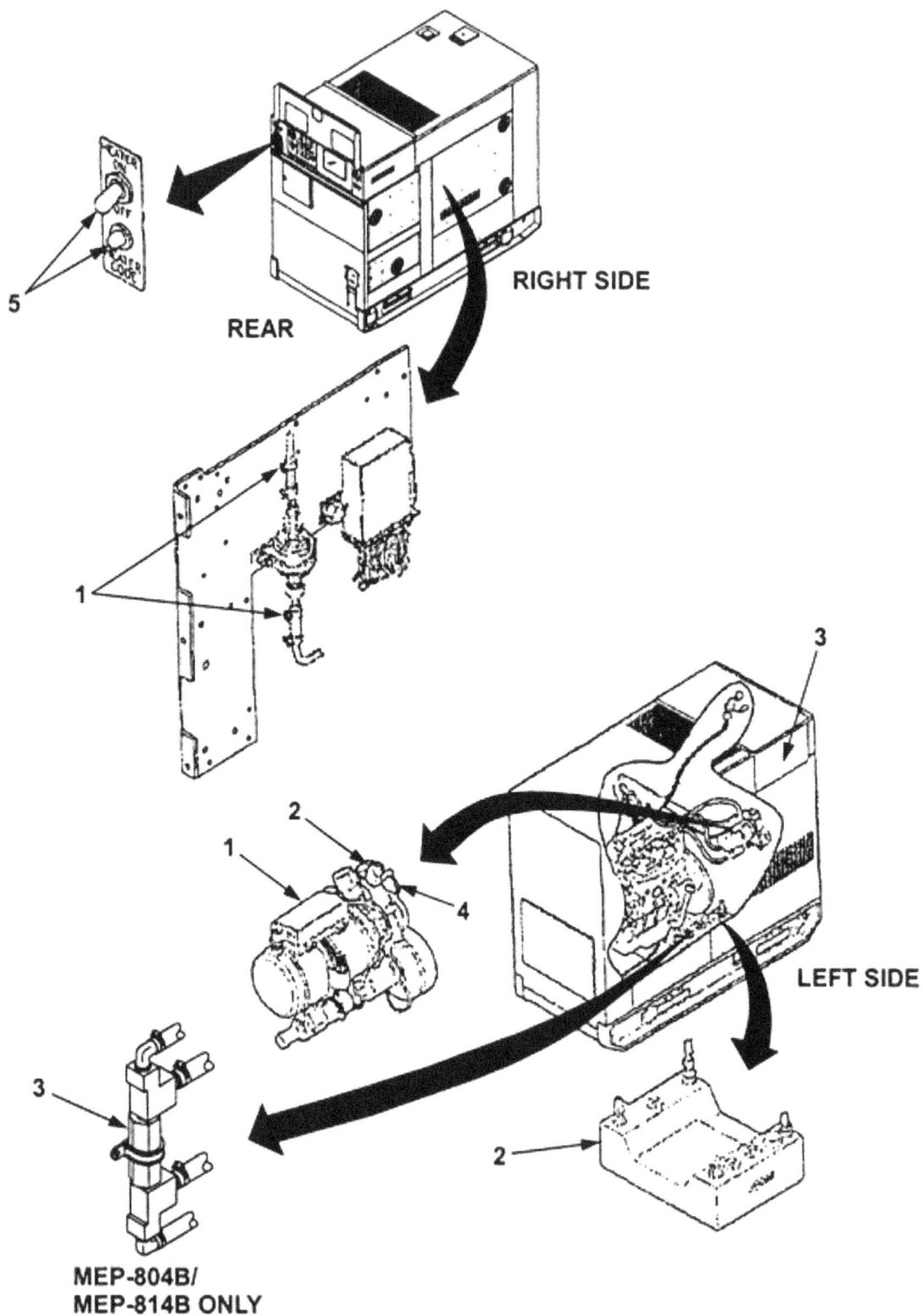

Figure 1. Operator PMCS Routing Diagram.

END OF WORK PACKAGE

OPERATOR MAINTENANCE

15 kW 50/60 AND 400 Hz SKID MOUNTED, TACTICAL QUIET GENERATOR SET

WINTERIZATION KIT, PREVENTIVE MAINTENANCE CHECKS AND SERVICES (PMCS)

Table 1. Operator Preventive Maintenance Checks and Services.

ITEM NO.	INTERVAL	ITEM TO BE CHECKED OR SERVICED	PROCEDURE	EQUIPMENT NOT READY/AVAILABLE IF:
		NOTE		
		Be sure that Generator Set PMCS is completed first in accordance with Chapter 4, WP 0011.		
1	Before	**VISUAL INSPECTION** HEATER ASSEMBLY	a. Check for damage.	Damage that renders equipment unsafe.
			b. Ensure that heater assembly is mounted securely.	Heater not mounted securely.
		CONTROL UNIT	Check for loose or broken wires or damage.	Wires loose or broken or control unit damaged.
		FUEL AND COOLANT LINES	Check on, around, and under equipment for fuel, oil, or coolant leaks.	Class III coolant or any class fuel leak is detected.
2	Before	HEATER	Inspect heater for signs of leaks.	~~Class III coolant or any~~ class fuel leak is detected.
		FUEL LINES	Inspect winterization kit fuel lines for kinks, leaks, loose or damaged clamps.	Fuel lines damaged; clamps missing. Any fuel leak.
		FUEL PUMP	Inspect fuel pump for leaks.	Hose obstructed; hose or clamp missing or damaged.
		EXHAUST HOSE	Inspect for obstruction, missing or damaged mounting clamp.	Inlet hose obstructed.
		AIR INLET HOSE	Inspect for obstruction, missing or damaged mounting clamp.	
		WARNING		
		Cooling system operates at high temperature and pressure. Contact with high pressure steam and/or liquids can result in burns and scalding. Shut down generator set, and allow system to cool before performing checks, services and maintenance. Failure to comply can cause injury or death to personnel.		
3	Before	WINTERIZATION KIT COOLANT LINES	Inspect for loose, damaged or missing clamps.	Class III leaks or missing clamps or hoses.
			Inspect for leaks.	Class III leaks or missing clamps or hoses.
		COOLANT PUMP	Inspect for leaks.	Class III leaks or missing clamps or hoses.
		BYPASS VALVE	Inspect for leaks, damage, loose clamps, or other damage.	Class III leaks or missing clamps or hoses.

Table 1. Operator Preventive Maintenance Checks and Services. - Continued

ITEM NO.	INTERVAL	ITEM TO BE CHECKED OR SERVICED	PROCEDURE	EQUIPMENT NOT READY/AVAILABLE IF:
4	Before	WIRE HARNESS	Inspect wiring for burned or frayedinsulation or loose terminals.	Wiring is loose or burned.
5	During	HEATER CONTROL AND SWITCH LAMP	a. Check that indicator light is on when heater is operating. b. Check Heater function Code Plate.	Lamp blinks showingfailure in accordance with Heater Function Code Plate.
6	After	HEATER ASSEMBLY CONTROL UNIT FUEL AND COOLANT LINES	a. Check for damage.b. Ensure that heater assembly is mounted securely. Loose or broken wires or damage. Check on, around, and under equipment for fuel, oil, or coolant leaks.	Damage that renderse-quipment unsafe. Heater not mountedse-curely. Wires loose or broken orcontrol unit damaged. Class III coolant or any-class fuel leak isdetected.
7	After	HEATER FUEL LINES FUEL PUMP EXHAUST HOSE AIR INLET HOSE	Inspect heater for signs of leaks. Inspect winterization kit fuel lines forkinks, leaks, loose or damaged clamps. Inspect fuel pump for leaks. Inspect for obstruction, missing or-damaged mounting clamp. Inspect for obstruction, missing or-damaged mounting clamp.	Fuel lines damaged,-clamps missing, or any leaks. Any fuel leak. Obstructed exhaust. Inlet hose obstructed.

WARNING

Cooling system operates at high temperature and pressure. Contact with high pressure steam and/orliquids can result in burns and scalding. Shut down generator set, and allow system to cool beforeperforming checks, services and maintenance. Failure to comply can cause injury or death topersonnel.

WARNING

Cooling system operates at high temperature and pressure. Contact with high pressure steam and/orliquids can result in burns and scalding. Shut down generator set, and allow system to cool beforeperforming checks, services and maintenance. Failure to comply can cause injury or death topersonnel.

8 After WINTERIZATION KIT

Inspect for loose, damaged, ormissing clamps.

Class III leaks or missingclamps or hoses.

COOLANT LINES

Inspect for leaks. Class III leaks or missing

clamps or hoses.

COOLANT PUMP Inspect for leaks. Class III leaks or missing

clamps or hoses.

Table 1. Operator Preventive Maintenance Checks and Services. - Continued

ITEM NO.	INTERVAL	ITEM TO BE CHECKED OR SERVICED	PROCEDURE	EQUIPMENT NOT READY/AVAILABLE IF:
		BYPASS VALVE	Inspect for leaks, damage, looseclamps, or other damage.	Class III leaks or missing-clamps or hoses.
9	After	WIRE HARNESS	Inspect wiring for burned or frayedinsulation or loose terminals.	Wiring is loose ordam-aged.
10	After	HEATER CONTROL AND SWITCH LAMP	Check that indicator light is operable. Check Heater Function Code Plate.	Indicator light not oper-able. Heater Function Code Plate missing.

END OF WORK PACKAGE

OPERATOR MAINTENANCE

15 kW 50/60 AND 400 Hz SKID MOUNTED, TACTICAL QUIET GENERATOR SET

WINTERIZATION KIT, MAINTENANCE PROCEDURES

MAINTENANCE

Refer to Chapter 4, Operator Maintenance Instructions for generator set maintenance procedures. Operator maintenance functions for the kit are limited to those described in WP 0011, Table 1, Operator Preventive Maintenance Checks and Services.

END OF WORK PACKAGE

INDEX

Subject WPSequenceNo.-PageNo.

INDEX

By Order of the Secretary of the Army:

Official:

JOYCE E. MORROW
Administrative Assistant to the
Secretary of the Army
322437

GEORGE W. CASEY, JR *General, United States Army*
Chief of Staff

By Order of the Secretary of the Air Force:

Official:

DONALD J. HOFFMAN
General, USAF
Command, AFMC

NORTON A. SCHWARTZ
General, USAF Chief of

Army Distribution:

To be distributed in accordance with the initial distribution number (IDN) 255265 requirements for TM 9-6115-643-10.

THE METRIC SYSTEM AND EQUIVALENTS

LINEAR MEASURE

1 Centimeter = 10 Millimeter = 0.01 Meters = 0.3937 Inches

1 Meter = 100 Centimeters = 1000 Millimeters = 39.37 Inches

1 Kilometer = 1000 Meters = 0.621 Miles

WEIGHTS

1 Gram = 0.001 Kilograms = 1000 Milligrams = 0.035 Ounces

1 Kilogram = 100 Grams = 2.2 lb.1 Cu. Meter = 1,000,000

1 Metric Ton = 1000 Kilograms = 1 Megagram = 1.1 Short Tons

LIQUID MEASURE

1 Millimeter = 0.001 Liters = 0.0338 Fluid Ounces

1 Liter = 1000 Millimeters = 33.82 Fluid Ounces

SQUARE MEASURE

1 Sq. Centimeter = 100 Sq. Millimeter = 0.155 Sq. Inches

1 Sq. Meter = 10,000 Sq. Centimeters = 10.76 Sq. Inches

1 Sq. Kilometer = 1,000,000 Sq. Meters = 0.386 Sq. Miles

CUBIC MEASURE

1 Cu. Centimeter = 1000 Cu. Millimeters = 0.06 Cu. Inches

1 Cu. Centimeters = 35.31 Cu. Feet

TEMPERATURE

5/9 (°F - 32) = °C

212° Fahrenheit is equivalent to 100° Celsius

90° Fahrenheit is equivalent to 32.2° Celsius

32° Fahrenheit is equivalent to 0° Celsius

9/5 °C + 32 = °F

APPROXIMATE CONVERSION FACTORS

TO CHANGE	TO	MULTIPLY BY
Inches	Centimeters	2.540
Feet	Meters	0.305
Yards	Meters	0.914
Miles	Kilometers	1.609
Square Inches	Square Centimeters	6.451
Square Feet	Square Meters	0.093
Square Yards	Square Meters	0.836
Square Miles	Square Kilometers	2.590
Acres	Square Hectometers	0.405
Cubic Feet	Cubic Meters	0.028
Cubic Yards	Cubic Meters	0.765
Fluid Ounces	Milliliters	29.573
Pints	Liters	0.473
Quarts	Liters	0.946
Gallons	Liters	3.785
Ounces	Grams	28.349
Pounds	Kilograms	0.454
Short Tons	Metric Tons	0.907
Pound-Feet	Newton-Meters	1.356
Pounds per Square Inch	Kilo pascals	6.895
Miles per Gallon	Kilometers per Liter	0.425
Miles per Hour	Kilometers per Hour	1.609

TO CHANGE	TO	DIVIDE BY
Centimeters	Inches	2.540
Meters	Feet	0.305
Meters	Yards	0.914
Kilometers	Miles	1.609
Square Centimeters	Square Inches	6.451
Square Meters	Square Feet	0.093
Square Meters	Square Yards	0.836
Square Kilometers	Square Miles	2.590
Square Hectometers	Acres	0.405
Cubic Meters	Cubic Feet	0.028
Cubic Meters	Cubic Yards	0.765
Milliliters	Fluid Ounces	29.573
Liters	Pints	0.473
Liters	Quarts	0.946
Liters-Meters	Gallons	3.785
Grams	Ounces	28.349
Kilograms	Pounds	0.454
Metric Tons	Short Tons	0.907
Newton-Meters	Pound-Feet	1.356
Kilo pascals	Pounds per Square Inch	6.895
Kilometers per Liter	Miles per Gallon	0.425
Kilometers per Hour	Miles per Hour	1.609

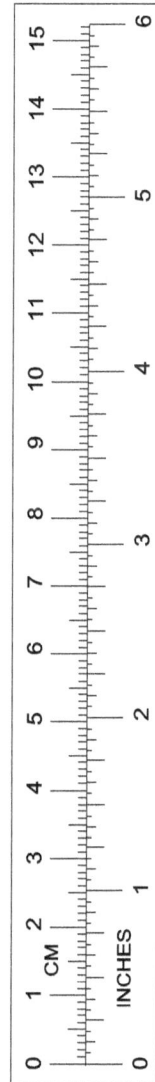